From Distributed Quantum Computing to Quantum Internet Computing

From Distributed Quantum Computing to Quantum Internet Computing

An Introduction

Seng W. Loke
Deakin University
Australia

WILEY

Published by John Wiley & Sons, Inc., Hoboken, New Jersey.
Published simultaneously in Canada.

For general information on our other products and services or for technical support, please contact our Customer Care Department within the United States at (800) 762-2974, outside the United States at (317) 572-3993 or fax (317) 572-4002.

Wiley also publishes its books in a variety of electronic formats. Some content that appears in print may not be available in electronic formats. For more information about Wiley products, visit our web site at www.wiley.com.

Library of Congress Cataloging-in-Publication Data Applied for:

[ISBN: 9781394185511; ePDF: 9781394185528; ePub: 9781394185535; oBook: 9781394185542]

Cover Design: Wiley
Cover Image: © Yuichiro Chino/Getty Images

Set in 9.5/12.5pt STIXTwoText by Straive, Chennai, India

To my Princess YC

Contents

About the Author

Professor Seng W. Loke (Member, IEEE) received the BSc degree (Hons.) in Computer Science from the Australian National University in 1994 and the PhD degree in Computer Science from the University of Melbourne, Australia, in 1998. He is a Professor of computer science at the School of Information Technology, Deakin University, Australia. He currently co-leads the IoT Platforms and Applications Laboratory and directs the Centre for Software, Systems and Society, within Deakin's School of Information Technology. His research interests include the Internet of Things, quantum Internet computing, cooperative vehicles, distributed and mobile systems, and smart cities.

Preface

The book is an outcome of my own journey, in recent years, into quantum computing and into a number of topics (detailed later), which can be considered to come under the theme of *quantum Internet computing*. This seems to be rather timely, since recent years have also seen important developments in quantum computing, including companies successfully building (though small/intermediate scale) quantum computers and offering access to them over the Cloud, and some momentum toward building the quantum Internet, making these topics almost too important to ignore. The developments in quantum computing, quantum cryptography, quantum information theory, quantum networking, and the quantum Internet in the recent decades have led to what has been called "the new quantum age" (to borrow the title from Andrew Whitaker's book) and the "second quantum revolution," which we are living in today.[1]

This book is an attempt to provide an introductory overview of work leading toward what I call *quantum Internet computing* at the intersection of work in distributed quantum computing and the quantum Internet, where one does distributed computing but over Internet-scale distances and systems involving nodes connected via the Internet. The notion of quantum Internet computing is based loosely on an analogy to *Internet computing* at the intersection of work in distributed computing and the Internet. While the quantum Internet and distributed quantum computing can be considered nascent, the book attempts a selective introduction to the following four key topics which I identify as coming under quantum Internet computing: (i) distributed quantum computing, including quantum protocols and theoretical perspectives, (ii) distributed quantum computing via nonlocal or distributed quantum gates, (iii) delegating

---

1 This phrase "second quantum revolution" has been used at the NIST website https://www .nist.gov/physics/introduction-new-quantum-revolution/second-quantum-revolution, and in association with developments which the contributions, recognized by the 2022 Nobel Prize in Physics, drove: https://www.nobelprize.org/uploads/2022/10/advanced-physicsprize2022.pdf [last accessed: 7/10/2022].

quantum computations, blind quantum computing, and verifying delegated quantum computations, and (iv) the quantum Internet, including the concept, and key ideas, such as quantum entanglement distillation, and entanglement swapping (where this book focuses on the computational aspects, instead of the underlying physics, if one accepts such a separation of focus). At the time of writing, distributed quantum computers over the quantum Internet can be considered "under construction," though experiments on fundamental concepts have been realized as well as prototypes of software working over simulations, and demonstrations of quantum networking – so, this book is rather optimistic toward the future.

The book, in a way, empathizes with newcomers to the subject and attempts to explain the concepts for the newcomer to quantum (Internet) computing, having been there myself. In short, the book is what I would have liked to have had when starting my own journey into these areas, at least to gain an overview of the area. Knowledge in the four topics above exists though mainly distributed in the research literature and in a range of (very good) online resources (which the book will also point out) – this book is an attempt (to the author's knowledge, the first book) to discuss the above four topics under one "umbrella paradigm," here called quantum Internet computing.

Each of the above four topics is emerging as active areas of research, and this book, as well, does not attempt a full coverage. The book aims to highlight and discuss a selection of key (at least in my opinion) ideas and concepts in the literature within those four topics above, and hence, the word "introduction" in the title, and given the vastness of the literature on the above topics, one might consider this introduction "short"! As a result, the book might appear to lack novelty and sophistication since it might not be a complete guide about the state of the art in all the above topics (multiple books would be likely needed to do so!), but the aim of this book is to introduce readers to certain key (and one might say fundamental) ideas that have emerged in the above topics, and perhaps "whet the appetite" of readers to go further. Even with the relatively "small" sampling of algorithms, protocols, and ideas discussed in this book, one can already observe their beauty and ingenuity – which is also partly a reason for the choice of topics included. One might consider the book as a "teaser" for a vast area of exciting research.

The book's approach is to select a range of key ideas and discuss them in depth, including detailed calculations often needed in seeing how a protocol or concept actually works – this is partly inspired by David McMahon's book *Quantum Computing Explained* which helps readers by detailing calculations. The details are given to allow someone fairly new to quantum computing (and not the expert who does such calculations every day!) to be able to see why certain ideas, algorithms, or protocols work; hence, it is an attempt to be pedagogical, providing a "gentle" introduction for those who might not have read the research papers upon which the book's material is based – we refer to and point out the original papers and resources for further reading.

The book is intended for advanced undergraduate and graduate students, researchers and practitioners in industry and academia, across different disciplines who might be interested in the area of quantum Internet computing and just wanted an introduction to the area, or before possibly moving on to technical papers and the research literature. However, given my own background in computing, and the style of the book, the book should appeal to students and researchers in Computer Science, or Information Technology (though it is hoped it might be useful to some physicists and mathematicians as well).

Often, the mathematics and quantum physics knowledge are hurdles to one, who does not have such background, trying to learn quantum computing for the first time. So, the book does not use a mathematical monograph style with theorems and proofs, but attempts a narrative style without sacrificing the mathematical rigour and technical details; in so doing, the theoretical Computer Scientist, physicist, or mathematician (or student thereof) might find the style of the book rather "informal," and there is little focus on exercises – but the aim is to make the book "readable" (perhaps an exercise for the reader is to follow the calculations and find errors (if any)!). The main prerequisites for readers would be mainly basic linear algebra and probability (and perhaps some perseverance to follow calculations!) and some basic background in computing, and though some basic knowledge in quantum computing would be helpful, we attempt to provide sufficient background for the assiduous reader relatively new to quantum computing (and provide pointers to good resources for learning quantum computing for the first time). The book can be used as a reference for a first course on quantum Internet computing (though perhaps not as a first course on quantum computing), supplemented by selected research literature.

The expert reader might notice that there are some topics missing, e.g. measurement-based quantum computation or one-way quantum computing, and quantum error correction (each of these topics probably deserves books of its own). The book also does not discuss distributed quantum sensing.

Additional explanations and calculations that could be useful to readers are in footnotes as well as marked as **Aside**. Lastly, the reader is invited to email me any inaccuracies or errors found in the book. With the above, let's get on with it!

October 2023

Seng W. Loke
Deakin University, Melbourne, Australia

Acknowledgments

The author would like to thank the School of Information Technology at Deakin University and the publisher (and the wonderful editorial team) for the tremendous support toward the writing of this book (any errors would be my own fault!). The quantum computing enthusiasts and colleagues at the School have also helped (implicitly) nudge me toward the subject and provided a vibrant environment for sustaining my interest in the subject. Finally, my gratitude to YC for her continual support.

Seng W. Loke

1

Introduction

> For the Stoics, living in accordance with nature required a knowledge of nature and its operations. One reason for this was that the study of nature was thought to offer the best way of establishing what lay within one's own power, and what in the power of nature.
>
> — Peter Harrison, The Territories of Science and Religion

1.1 The New Quantum Age and the Second Quantum Revolution

Andrew Whitaker's book *The New Quantum Age: From Bell's Theorem to Quantum Computation and Teleportation* speaks of the First Quantum Age marked by the pioneers of *quantum theory* (or *quantum mechanics*), such as the physicists Max Planck, Albert Einstein, Niels Bohr, and others, in the first quarter of the 20th century, and the New Quantum Age ushered in by the work of the physicist John Stewart Bell in the 1960s and others, some later winning the Nobel Prize in Physics, in the area of *quantum information science* (a term we will come back to later).

The 2022 Nobel Prize in Physics went to three outstanding physicists, Alain Aspect, John F. Clauser, and Anton Zeilinger, for "experiments with entangled photons, establishing the violation of Bell inequalities and pioneering quantum information science."[1] Some readers might already be familiar with the many ideas and concepts mentioned in that one sentence, but some not so – we will in this book unpack some of the above concepts such as "entanglement," "quantum information," and "Bell inequalities" (and the associated concepts of "Bell pairs"

---

1 See https://www.nobelprize.org/prizes/physics/2022/press-release [last accessed: 7/10/2022].

From Distributed Quantum Computing to Quantum Internet Computing: An Introduction,
First Edition. Seng W. Loke.

and "Bell states," named after John Bell mentioned above). Some of the experiments conducted by the Nobel prize winners investigated and demonstrated a key concept in quantum mechanics called *entanglement*, a term due to physicist Erwin Schrödinger in 1935[2] – informally, a type of "quantum link."[3] One could also think of such entanglement between particles (e.g. photons of light) as a type of "resource" that will enable the transfer of quantum information over geographical distances (a phenomenon called *quantum teleportation*). We will see that such a resource is central to distributed quantum computing and is a key concept in quantum networks.

As mentioned in the document on the scientific contributions of the 2022 Nobel Prize in Physics:

> This year's Nobel prize is for experimental work. Apart from the disparities in philosophical interpretation, the early Bell experiments drove the development of what is often referred to as the 'Second Quantum Revolution'. Two of this year's laureates, John Clauser and Alain Aspect, are honoured for work that initiated a new era, opening the eyes of the physics community to the importance of entanglement, and providing techniques for creating, processing and measuring Bell pairs in ever more complex and mind-boggling scenarios. The experimental work of the third Laureate, Anton Zeilinger, stands out for its innovative use of entanglement and Bell pairs, both in curiosity driven fundamental research and in applications such as quantum cryptography. [https://www.nobelprize.org/uploads/ 2022/10/advanced-physicsprize2022.pdf, p. 15, accessed: 7/10/2022]

The Nobel Prize also recognized Anton Zeilinger's work on entanglement swapping and multipartite entanglement, concepts which we will discuss later in the book.

Since the early experimental work by the Nobel prize winners dating back to the early 1970s and 1980s, much has happened in the areas of experimental demonstrations of quantum entanglement (over larger geographical scales) and quantum teleportation, quantum cryptographic protocols, quantum communications, quantum networking, quantum computing, quantum distributed computing, as well as quantum information theory. For example, in 1984, Charles Bennett and Gilles Brassard came up with the first quantum key

---

2 See https://plato.stanford.edu/entries/qt-entangle [last accessed: 8/10/2022].
3 The word "link" is likely not an accurate description of quantum entanglement, but suffices here to paint a mental picture of some kind of connection between particles for the purposes of this chapter. Einstein associated it with "spooky action at a distance," which John Bell's insight and the Nobel Prize winners' experiments showed does occur in reality. We discuss entanglement in detail later in the book.

distribution (QKD) protocol, a secure way to share keys used for encryption and decryption, called BB84,[4] which was later demonstrated experimentally in 1989 [Bennett et al., 1992]. A brief history of quantum cryptography is given by Brassard [2005]. In 1991, Artur Ekert came up with the E91 protocol for QKD [Ekert, 1991]. We will come back to the topic of QKD later in the book. Further experimental demonstrations were then conducted over the years. In the early 2000s, ID Quantique[5] became one of the first companies to bring a QKD product to the commercial market. Going beyond just two nodes, the world's first *quantum network* became operational between 2004 and 2007, demonstrating QKD, i.e. the DARPA Quantum Network.[6] Today, a Quantum Network architecture standard is being developed with the creation of the Quantum Internet Research Group (an Internet Research Task Force [IRTF]).[7] Recent work has continued to conceptualize and develop architectures and applications of the quantum Internet, as reviewed in Gyongyosi and Imre [2022], Illiano et al. [2022], Wehner et al. [2018], and Rohde [2021]. Quantum-enabled 6G wireless networking has been discussed in Wang and Rahman [2022]. We discuss quantum networking and the quantum Internet further in Chapter 6.

At the same time, developments in quantum computing have progressed with (i) important work in the 1980s and 1990s, e.g. with the foundational thinking of Deutsch [1985] and the invention of quantum algorithms for factoring numbers by Shor [1994, 1999] and for quantum search by Grover [1996] and (ii) many other developments in the more recent decades, including in the areas of quantum computing applications such as quantum simulation [Smith et al., 2019] and quantum machine learning [Biamonte et al., 2017; Ramezani et al., 2020; Schuld and Petruccione, 2021]. Several companies (big tech and startups)[8] and a number of universities have built quantum computers or are experimenting with quantum hardware concepts,[9] research continues into building even larger scale quantum computers, and developing tools and software for programming quantum computers toward full quantum computer systems [Ding and Chong, 2020] for at least, what John Preskill has called, *Noisy Intermediate-Scale Quantum (NISQ)*

---

4 The 1984 published paper entitled "QUANTUM CRYPTOGRAPHY: PUBLIC KEY DISTRIBUTION AND COIN TOSSING" by C. Bennett and G. Brassard is online at https://arxiv.org/pdf/2003.06557.pdf [last accessed: 9/10/2022].
5 See company Website: https://www.idquantique.com [last accessed: 16/10/2022].
6 See https://apps.dtic.mil/dtic/tr/fulltext/u2/a471450.pdf and https://arxiv.org/pdf/quant-ph/0412029.pdf [last accessed: 2/8/2022].
7 See https://irtf.org/qirg and in particular, https://datatracker.ietf.org/doc/rfc9340 [last accessed: 14/4/2023].
8 See a (relatively long) list at https://thequantuminsider.com/2022/09/05/quantum-computing-companies-ultimate-list-for-2022 [last accessed: 16/10/2022].
9 See a list at https://thequantuminsider.com/2022/05/16/quantum-research [last accessed: 16/10/2022].

computers [Preskill, 2018].[10] Also emerged is what has been called *quantum software engineering* [Piattini and Murillo, 2022; De Stefano et al., 2022; Ali et al., 2022], concerned with processes, tools, and methodologies for developing software that runs on quantum computer systems. A number of companies are also providing access to quantum computers via a cloud service model.[11] Government investments into quantum computing and networking have increased in many countries.[12]

Hence, one can see the increasing developments at the intersection of Information and Communication Technologies (ICT) and quantum theory, yielding *quantum information science*, which can be described as

> an emerging field with the potential to cause revolutionary advances in fields of science and engineering involving computation, communication, precision measurement, and fundamental quantum science. [https://www.nsf.gov/pubs/2000/nsf00101/nsf00101.htm, accessed: 8/10/2022]

And more recently, research into quantum software development, from the information technology or computing perspective.

1.2 Distributed Quantum Computing and the Rise of Quantum Internet Computing

We have been in the Internet or Web Age for some decades now since the early days of the Web in the 1990s and the invention of email even earlier. With networked computers (wired or wireless) around the world, and computers being pervasive, we then have the field of *distributed computing*, looking into computations (and communication protocols) over distributed networked or connected nodes, and the *Internet of Things*, which is concerned with all sorts of things (including everyday objects with embedded computers), people and places, becoming connected

---

10 A *qubit* refers to the basic unit of quantum information and NISQ computers refers to computers having quantum processors with 50 to a few hundred qubits, where the number of qubits in a computer indicates the computational capability of the computer – the more qubits the more *quantum parallelism* possible, and hence, the more operations that can be performed (concurrently) per unit of time.
11 It is interesting that many companies (e.g. Google [https://quantumai.google/cirq/google/concepts], Rigetti [https://qcs.rigetti.com/sign-in], IBM [https://quantum-computing.ibm.com], Amazon [https://aws.amazon.com/braket] and so on; a more extensive list is: https://thequantuminsider.com/2022/05/03/13-companies-offering-quantum-cloud-computing-services-in-2022), provide usage of quantum computers via a cloud computing service model.
12 See a 2021 list of government investments into quantum computing at https://www.insidequantumtechnology.com/news-archive/government-investments-in-quantum-computing-around-the-globe [last accessed: 16/10/2022].

to the Internet. Distributed computing might be over computers (or nodes) within the same room, or the same geographical area, or might utilize nodes geographically far apart but connected over the Internet. The latter involves a number of issues perhaps not as apparent or on the same scale as when the nodes are local, such as increased latency in communications, fault tolerance, heterogeneity, and scalability. Distributed computing can go beyond geographically local nodes and involve nodes distributed over vast geographical (Internet size) scales (perhaps even interplanetary in the future!), and hence, the often used term *Internet computing* in such cases.

1.2.1 Distributed Quantum Computing

While work on distributed computing (including mobile and pervasive computing) in the recent decades have been mostly on classical (one might call traditional) distributed computing, there has been a lot of thinking since the 1990s about how quantum mechanics might have an impact on distributed computing.

Lov Grover proposed the idea of computations with distributed quantum processors [Grover, 1997], which he called *quantum telecomputation*, with respect to the problem of finding the average of N real numbers to a given precision, where each node has one or more pieces of the data.

Others studied the communication complexity of quantum distributed system protocols, where some computation is to be computed by two or more parties which need to send messages to each other, including how having entanglement between nodes might help reduce the communications required between the nodes, compared with classical versions of the distributed system protocols, and how entanglement might make possible some distributed computations not possible classically, e.g. the work by Buhrman and Röhrig [2003], Buhrman et al. [2010], Broadbent and Tapp [2008], and Cleve and Buhrman [1997] – such work has been called *distributed quantum computing*, dating back to late 1990s and early 2000s. Cirac et al. [1999] studied considerations of noisy channels on distributed quantum computations.

At around the same time, there has been work on non-local quantum gates (viewing quantum gates as the quantum analogue of digital logic gates), where quantum computations are distributed over two or more nodes [Yimsiriwattana and Lomonaco, 2005; Eisert et al., 2000], including a distributed version of Shor's algorithm mentioned earlier [Yimsiriwattana and Lomonaco, 2004], also called *distributed quantum computing*.

Since then, more recent work on what might be considered as "(classical) distributed computing inspired" distributed quantum computing have taken place, e.g. the work by Parekh et al. [2021], Sundaram et al. [2022], Denchev and Pandurangan [2008], Andrés-Martínez and Heunen [2019], and Häner et al. [2021], many of which involve work on distributing quantum computations (represented as *quantum circuits*, analogous to digital circuits) over multiple nodes.

Indeed, apart from theoretical studies into quantum communication complexity, a key motivation for distributed quantum computing is the ability to utilize multiple quantum computers for a given application:

> By connecting a network of limited capacity quantum computers via classical and quantum channels, a group of small quantum computers can simulate a quantum computer with a large number of qubits. This approach is useful for the development of quantum computers because the earliest useful quantum computers will most likely hold only a small number of qubits. [Yimsiriwattana and Lomonaco, 2005, p. 131]

Of course, what "small" meant at that time might be different today, or in the NISQ era.

For a collection of quantum computers to collaboratively perform computations, the computers might not just exchange classical information but also quantum information. Connecting multiple quantum computers together to perform computations is non-trivial and requires not just classical communication (via classical channels) between these computers but also quantum communication (via quantum channels) between these computers so that quantum information can be transferred between these computers, not just classical information. In particular, we will see that two (or more) computers can use shared entanglement in order to exchange quantum information. An excellent depiction of such a connected set of quantum computers (or one might say *quantum processing units* (QPUs) perhaps analogous to GPUs) is given in DiAdamo et al. [2021] and redrawn slightly differently in Figure 1.1. One can observe that both types of networks, classical and quantum, are required and when the computers are connected not just via a local network but the Internet, the classical network becomes the classical Internet and the quantum network becomes the quantum Internet. Also needed is a way to manage the programs that will run across multiple computers, such as scheduling the (sub)programs on each computer, managing the program execution and communications, and coordinating and combining the results and so on, i.e. some sort of controller is required though decentralized controllers are also discussed in DiAdamo et al. [2021]. There would be greater complexity if resources (e.g. QPUs, entanglement generation, and network access) are shared by multiple programs and clients wanting to execute their programs.

We note that there has also been work on the quantum parallel RAM model, analogous to the classical parallel RAM model for representing parallel and distributed computations [Beals et al., 2013], and work on quantum arithmetic algorithms running on a distributed quantum computer (called a quantum

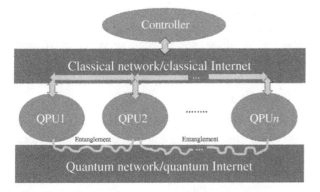

Figure 1.1 An illustration of a distributed quantum computer with multiple quantum computers (or *quantum processing units*, or QPUs) connected using quantum channels (i.e. via entanglement) and classical channels, so that both quantum and classical information can be communicated. The controller manages the execution of computations, classical and quantum parts, across the computers (QPUs). Source: Adapted from the diagram by DiAdamo et al. [2021].

multicomputer) in Van Meter et al. [2008]. Based on the measurement-based model of quantum computing by Raussendorf et al. [2003], a model for *distributed measurement-based quantum computations* has been developed [Danos et al., 2007]. A formal model for reasoning about distributed quantum computing has been developed by Ying and Feng [2009]. Interestingly, the quantum Message Passing Interface [Häner et al., 2021] provides an Application Programming Interface (API) for distributed quantum computing, analogous to the Message Passing Interface (MPI) developed as a standardized API for (classical) distributed computing [Nielsen, 2016]; by writing programs using such a standardized API, such programs become portable, and can, by and large, be independent of the underlying computer architecture or specific implementations.

There have also been studies on "quantum versions" of classical distributed computing protocols, though fundamentally, these are quite different from the classical analogues since quantum entanglement is employed, including quantum oblivious transfer, QKD, quantum coin flipping, quantum electronic voting, quantum leader election, quantum anonymous broadcasting, quantum secret sharing, and quantum Byzantine Generals.[13] We will look at some of these protocols later in the book.

Hence, drawing inspiration partly from the decades of work in (classical) distributed computing and other innovations, developments have emerged in distributed quantum computing, including new concepts, models, abstractions,

---

13 See a list at https://wiki.veriqloud.fr/index.php?title=Protocol_Library [last accessed: 16/10/2022].

protocols, tools and software. We have not provided a comprehensive account of distributed quantum computing here, but a brief overview, and we will come back to many of the topics above in the rest of the book. A recent survey on distributed quantum computing is by Caleffi et al. [2022].

1.2.2 Quantum Internet Computing

With developments in the coming decades in quantum networking (and the quantum Internet), quantum computing hardware and software, and distributed quantum computing, we can conceptualise *quantum Internet computing*, analogous to (classical) Internet computing mentioned above, referring to distributed quantum computing over the emerging quantum Internet, potentially over vast geographical distances, or roughly put:

$$\begin{matrix} quantum \\ Internet \\ computing \end{matrix} = \begin{matrix} distributed \\ quantum \\ computing \end{matrix} + \begin{matrix} quantum \\ Internet \end{matrix}$$

Inspired by Cuomo et al. [2020] who noted that the quantum Internet will become the "underlying infrastructure of the Distributed Quantum Computing ecosystem," Figure 1.2 illustrates a layered conceptualization of the idea of distributed quantum computing over the quantum Internet, where

i) the Distributed Quantum Computing Layer contains the
   - the Distributed Quantum Computing Applications sublayer, which comprises the end-user applications built using the tools and abstractions from the Distributed Quantum Computing Tools, Abstractions, Libraries, Environment layer, and
   - the Distributed Quantum Computing Tools, Abstractions, Libraries, Environment sublayer, which provides a range of tools and abstractions, and environments for executing and managing programs.

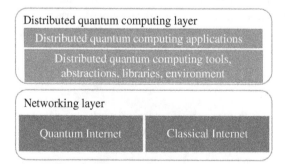

Figure 1.2 A simple architecture of quantum Internet computing: distributed quantum computing over the quantum Internet.

ii) the Networking Layer realizes quantum networking services used by the programs at the Distributed Quantum Computing Layer, supported by the classical Internet, including preparation of quantum entanglement and transmission of qubits; at least for the foreseeable future, the quantum Internet is expected to coexist with the classical Internet.

Later in the book, we will come back to discussing this architecture in further detail, but here, it suffices to convey the idea of quantum Internet computing.

1.3 Aim and Scope of the Book

As mentioned, Internet computing, where one does distributed computing but over Internet scale distances and distributed systems involve nodes connected via the Internet, is at the intersection of work in (classical) distributed computing and the (classical) Internet. By an analogy to Internet computing, one could ask the question of what would be at the intersection of distributed quantum computing and the quantum Internet. This book provides an introductory overview of selected topics in distributed quantum computing and the quantum Internet, and then attempts to answer this question, proposing the notion of quantum Internet computing, a proposed "umbrella paradigm" for a collection of topics (listed below), from an analogy to (classical) Internet computing:

- distributed quantum computing, including theoretical studies of quantum protocols which are often quantum versions of classical quantum distributed computing protocols, and perspectives on communication complexity including gains from the use of entanglement,
- distributed quantum computing using non-local (or distributed) quantum gates and circuits, where we focus on how large quantum computation represented by a quantum circuit can be distributed over multiple nodes,
- delegating quantum computations to a server, including delegating in such a way that the server has limited or no knowledge of the computations for the purposes of privacy of the client (who is delegating the computations to the server) or sometimes called *blind quantum computing*, and verifying delegated quantum computations so that the client is assured that the server did do the required computations; delegating quantum computations (on a pay-per-use basis) might come under what has been called *quantum cloud computing* given the cloud model of providing computing-as-a-service (typically, in classical cloud computing, including software, computer servers, memory and so on, provided on a pay-as-you-use service over the Internet), and
- quantum Internet, where we introduce the concept, key ideas such as quantum entanglement distillation, and entanglement swapping, and possible architectures.

With respect to Figure 1.2, the first three topics might be considered as falling within the Distributed Quantum Computing Layer, while the last topic, clearly, is in the Networking layer. In a way, "quantum Internet computing" is yet another label, but the label aims to (i) capture the scale of distributed quantum computing envisioned, and the associated challenges with that, as well as (ii) attempts to "bundle" the topics above with the issues they address, at least for the purposes of this book. There are a number of topics mentioned in Section 1.2.1, which we do not cover in detail in this book – some are covered in the books we will mention later in Section 1.5 and others are perhaps left to other books to come! The range of topics under quantum Internet computing may yet grow and evolve in the coming years.

1.4 Outline of this Book

The rest of this book is organized as follows.

Chapter 2 discusses the mathematical and quantum computing background required for the rest of the book, including a brief introduction to quantum computing and related concepts, including qubit and qubit states, measurement, quantum gates and circuits, entanglement, and teleportation.

Chapter 3 discusses a selection of distributed quantum computing problems and quantum protocols, many which have their origins in classical distributed computing. The chapter first discusses the distributed three-party product problem, the distributed Deutsch–Jozsa promise problem, and the distributed intersection problem, demonstrating how, for some problems, quantum information can enable fewer bits of communication to be used for a solution. We also look at how certain distributed computation problems can be solved with quantum information, but not possible classically. The second part of the chapter discusses quantum protocols, including quantum coin flipping, quantum leader election, QKD, quantum anonymous broadcasting, quantum voting, quantum Byzantine Generals, quantum secret sharing, and quantum oblivious transfer.

Chapter 4 discusses non-local gates, or quantum gates, for computations over several nodes. The chapter discusses the distributed CNOT gate, different ways for its implementation, and distributed control of gates beyond the CNOT gate. The chapter also briefly reviews work on distributed quantum circuits, distributed quantum computing algorithms using such non-local gates, and control and management for distributed quantum computers.

Chapter 5 discusses the idea of a client, with limited resources, delegating quantum computations to a more powerful quantum computer or server, in the style of what we see with cloud computing today. However, security and privacy are key considerations when delegating computations. We look at how to delegate quantum computations to a quantum server in such a way as to protect the privacy of the client's inputs, that is, the server cannot know the client's inputs while still being able to perform the computations for the client. The chapter looks at how the client can verify the server's computations, or at least detect if the server really performed the delegated computations correctly. The chapter also looks briefly at the idea of quantum computing as a service (QaaS).

Chapter 6 discusses some of the key concepts and ideas of the quantum Internet, including the central role that entanglement plays, quantum Internet architecture, and the concepts of entanglement swapping and entanglement purification, and briefly reviews the notion of quantum repeaters.

Chapter 7 brings the whole book together by connecting the topics discussed in the previous chapters and explaining how they would underlie the emerging area of quantum Internet computing, complementary to classical Internet computing.

1.5   Related Books and Resources

For readers interested in learning further, we mention books and Web resources specifically related to the topics of this book, i.e. the quantum Internet, and mention briefly how they are different from but complement this book.

This book is different from other excellent books on the quantum Internet such as

- Rodney van Meter's *Quantum Networking* (2014), Wiley, which is likely the earliest (or one of the earliest) books on the quantum Internet, captures a range of concepts and ideas which we will go in depth in Chapter 6 of this book; this book does not cover all the topics on the quantum Internet covered in van Meter's book but, for some of the concepts and ideas, we go deeper, including calculations.
- Peter Rohde's *The Quantum Internet: The Second Quantum Revolution* (2021), Cambridge University Press (CUP), which provides an excellent up-to-date comprehensive overview of the developments in the quantum Internet, covering many of the topics mentioned in this proposed book, but different from Rohde's book, this book will also cover distributed quantum computing and goes in depth into the calculations, in a pedagogical style, for some of the key protocols and concepts.

And there have been several books on the quantum Internet intended for general non-technical readers (which differs from this book which is for the technical reader) such as

- (The late) Jonathan Dowling's *Schrödinger's Web: Race to Build the Quantum Internet* (2020), CRC Press, which provides an excellent overview of related Physics concepts and ideas, and how the quantum Internet might develop in the years to come.
- Gösta Fürnkranz's *The Quantum Internet: Ultrafast and Safe from Hackers* (2020), Springer, which is also a readable introduction to the background in Physics and concepts for the quantum Internet, and also discusses recent developments.

This book focuses on "high-level" quantum distributed computing and protocols, but there are excellent books on quantum communications and quantum information theory such as

- Gianfranco Cariolaro's *Quantum Communications* (2015), Springer, which is one of the earliest (if not the earliest) book on quantum communications, on the differences between classical and quantum communications and foundational theory from a telecommunications perspective.
- Riccardo Bassoli, Holger Boche, Christian Deppe, Roberto Ferrara, Frank H. P. Fitzek, Gisbert Janssen, and Sajad Saeedinaeeni's *Quantum Communication Networks* (2021), Springer, which provides a recent tutorial on quantum networking, quantum information theory, quantum computing, and quantum network simulation tools.
- Ivan Djordjevic's *Quantum Communication, Quantum Networks, and Quantum Sensing* (2022), Elsevier, which provides an introduction to quantum communication, quantum error correction, quantum networks, and quantum sensing.

While this book will discuss QKD protocols, among many other topics, other books have focused primarily on quantum cryptography, in particular, with extensive discussions focusing on QKD and quantum key agreement (QKA) such as

- Ramona Wolf's *Quantum Key Distribution: An Introduction with Exercises* (2021), Springer, which is an excellent introduction to the large area of QKD with mathematical and technical depth.
- Federico Grasselli's *Quantum Cryptography: From Key Distribution to Conference Key Agreement* (2021), Springer, which covers QKD and QKA in a mathematically rigorous way.
- Miralem Mehic, Stefan Rass, Peppino Fazio, and Miroslav Voznak's book *Quantum Key Distribution Networks: A Quality of Service Perspective* (2022), Springer, which discusses practical aspects of deploying QKD over the Internet.

For those interested in quantum information theory, four books are:

- Benjamin Schumacher (who coined the term *qubit* in 1995) and Michael Westmoreland's *Quantum Processes Systems, and Information* (2010), CUP, has written an excellent book introducing quantum theory and the quantum information perspective.

- Mark M. Wilde's *Quantum Information Theory* (2013), CUP, might be considered a "bible" for quantum information theory! Mark also has notes available online entitled *From Classical to Quantum Shannon Theory* at https://markwilde.com/qit-notes.pdf [last accessed: 18/10/2022].
- Mahahito Hayashi's *Quantum Information Theory: Mathematical Foundation (2nd edition)* (2017), Springer, for the mathematically inclined.
- Joseph M. Renes' *Quantum Information Theory: Concepts and Methods* (2022), De Gruyter, provides an excellent introduction to quantum information theory.

Two particular Web resources have been very helpful in my own journey in learning about quantum networking and the quantum Internet:

- Keio University's Quantum Academy Of Science And Technology provides an excellent video series on introductory quantum networking and the quantum Internet, at https://www.youtube.com/c/QuantumCommEdu [last accessed: 18/10/2022].
- QuTech has an excellent site *Quantum Internet Explorer* for anyone interested in learning about the quantum Internet, at https://www.quantum-network.com [last accessed: 18/10/2022].

References

Shaukat Ali, Tao Yue, and Rui Abreu. When software engineering meets quantum computing. *Communications of the ACM*, 65(4):84–88, Mar 2022. ISSN 0001-0782. doi: 10.1145/3512340.

Pablo Andrés-Martínez and Chris Heunen. Automated distribution of quantum circuits via hypergraph partitioning. *Physical Review A*, 100:032308, Sep 2019. doi: 10.1103/PhysRevA.100.032308. URL https://link.aps.org/doi/10.1103/PhysRevA.100.032308.

Robert Beals, Stephen Brierley, Oliver Gray, Aram W. Harrow, Samuel Kutin, Noah Linden, Dan Shepherd, and Mark Stather. Efficient distributed quantum computing. *Proceedings of the Royal Society A: Mathematical, Physical and Engineering Sciences*, 469(2153):20120686, 2013.

Charles H. Bennett, François Bessette, Gilles Brassard, Louis Salvail, and John Smolin. Experimental quantum cryptography. *Journal of Cryptology*, 5(1):3–28, 1992.

Jacob Biamonte, Peter Wittek, Nicola Pancotti, Patrick Rebentrost, Nathan Wiebe, and Seth Lloyd. Quantum machine learning. *Nature*, 549(7671):195–202, 2017.

G. Brassard. Brief history of quantum cryptography: a personal perspective. In *IEEE Information Theory Workshop on Theory and Practice in Information-Theoretic Security, 2005*, pages 19–23, 2005. doi: 10.1109/ITWTPI.2005.1543949.

Anne Broadbent and Alain Tapp. Can quantum mechanics help distributed computing? *SIGACT News*, 39(3):67–76, Sep 2008. ISSN 0163-5700. doi: 10.1145/1412700.1412717.

Harry Buhrman and Hein Röhrig. Distributed quantum computing. In Branislav Rovan and Peter Vojtáš, editors, *Mathematical Foundations of Computer Science 2003*, pages 1–20, Springer-Verlag, Berlin, Heidelberg, 2003.

Harry Buhrman, Richard Cleve, Serge Massar, and Ronald de Wolf. Nonlocality and communication complexity. *Reviews of Modern Physics*, 82:665–698, Mar 2010. doi: 10.1103/RevModPhys.82.665.

Marcello Caleffi, Michele Amoretti, Davide Ferrari, Daniele Cuomo, Jessica Illiano, Antonio Manzalini, and Angela Sara Cacciapuoti. Distributed quantum computing: a survey. *CoRR*, Dec 2022. URL https://doi.org/10.48550/arXiv.2212.10609.

J. I. Cirac, A. K. Ekert, S. F. Huelga, and C. Macchiavello. Distributed quantum computation over noisy channels. *Physical Review A*, 59:4249–4254, Jun 1999. doi: 10.1103/PhysRevA.59.4249.

Richard Cleve and Harry Buhrman. Substituting quantum entanglement for communication. *Physical Review A*, 56:1201–1204, Aug 1997. doi: 10.1103/PhysRevA.56.1201.

Daniele Cuomo, Marcello Caleffi, and Angela Sara Cacciapuoti. Towards a distributed quantum computing ecosystem. *IET Quantum Communication*, 1(1):3–8, 2020.

Vincent Danos, Ellie D'Hondt, Elham Kashefi, and Prakash Panangaden. Distributed measurement-based quantum computation. *Electronic Notes in Theoretical Computer Science*, 170:73–94, 2007.

Manuel De Stefano, Fabiano Pecorelli, Dario Di Nucci, Fabio Palomba, and Andrea De Lucia. Software engineering for quantum programming: how far are we? *Journal of Systems and Software*, 190:111326, 2022.

Vasil S. Denchev and Gopal Pandurangan. Distributed quantum computing: a new frontier in distributed systems or science fiction? *SIGACT News*, 39(3):77–95, Sep 2008. ISSN 0163-5700. doi: 10.1145/1412700.1412718.

D. Deutsch. Quantum theory, the Church–Turing principle and the universal quantum computer. *Proceedings of the Royal Society of London Series A*, 400(1818):97–117, Jul 1985. doi: 10.1098/rspa.1985.0070.

Stephen DiAdamo, Marco Ghibaudi, and James Cruise. Distributed quantum computing and network control for accelerated vqe. *IEEE Transactions on Quantum Engineering*, 2:1–21, 2021. doi: 10.1109/TQE.2021.3057908.

Yongshan Ding and Frederic T. Chong. Quantum computer systems: research for noisy intermediate-scale quantum computers. *Synthesis Lectures on Computer Architecture*, 15(2):1–227, 2020.

J. Eisert, K. Jacobs, P. Papadopoulos, and M. B. Plenio. Optimal local implementation of nonlocal quantum gates. *Physical Review A*, 62:052317, Oct 2000. doi: 10.1103/PhysRevA.62.052317.

Artur K. Ekert. Quantum cryptography based on Bell's theorem. *Physical Review Letters*, 67:661–663, Aug 1991. doi: 10.1103/PhysRevLett.67.661.

Lov K. Grover. A fast quantum mechanical algorithm for database search. In *Proceedings of the Twenty-Eighth Annual ACM Symposium on Theory of Computing*, STOC '96, pages 212–219, Association for Computing Machinery, New York, NY, USA, 1996. ISBN 0897917855. doi: 10.1145/237814.237866.

Lov K. Grover. Quantum Telecomputation. *arXiv e-prints*, art. quant-ph/9704012, April 1997.

Laszlo Gyongyosi and Sandor Imre. Advances in the quantum internet. *Communications of the ACM*, 65(8):52–63, Jul 2022. ISSN 0001-0782. doi: 10.1145/3524455.

Thomas Häner, Damian S. Steiger, Torsten Hoefler, and Matthias Troyer. Distributed quantum computing with QMPI. In *Proceedings of the International Conference for High Performance Computing, Networking, Storage and Analysis*, SC '21, Association for Computing Machinery, New York, NY, USA, 2021. ISBN 9781450384421. doi: 10.1145/3458817.3476172.

Jessica Illiano, Marcello Caleffi, Antonio Manzalini, and Angela Sara Cacciapuoti. Quantum internet protocol stack: a comprehensive survey. *Computer Networks*, 213:109092, 2022.

Frank Nielsen. *Introduction to MPI: The Message Passing Interface*, pages 21–62. 2016. ISBN 978-3-319-21902-8. doi: 10.1007/978-3-319-21903-5_2.

R. Parekh, A. Ricciardi, A. Darwish, and S. DiAdamo. Quantum algorithms and simulation for parallel and distributed quantum computing. In *2021 IEEE/ACM Second International Workshop on Quantum Computing Software (QCS)*, pages 9–19, IEEE Computer Society, Los Alamitos, CA, USA, Nov 2021. doi: 10.1109/QCS54837.2021.00005. URL https://doi.ieeecomputersociety.org/10.1109/QCS54837.2021.00005.

Mario Piattini and Juan Manuel Murillo. *Quantum Software Engineering Landscape and Challenges*, pages 25–38. Springer International Publishing, Cham, 2022.

John Preskill. Quantum Computing in the NISQ era and beyond. *Quantum*, 2:79, Aug 2018. ISSN 2521-327X. doi: 10.22331/q-2018-08-06-79.

Somayeh Bakhtiari Ramezani, Alexander Sommers, Harish Kumar Manchukonda, Shahram Rahimi, and Amin Amirlatifi. Machine learning algorithms in quantum computing: a survey. In *2020 International Joint Conference on Neural Networks (IJCNN)*, pages 1–8, 2020. doi: 10.1109/IJCNN48605.2020.9207714.

Robert Raussendorf, Daniel E. Browne, and Hans J. Briegel. Measurement-based quantum computation on cluster states. *Physical Review A*, 68:022312, Aug 2003. doi: 10.1103/PhysRevA.68.022312.

Peter P. Rohde. *The Quantum Internet: The Second Quantum Revolution*. Cambridge University Press, 2021. doi: 10.1017/9781108868815.

M. Schuld and F. Petruccione. *Machine Learning with Quantum Computers*. Quantum Science and Technology. Springer International Publishing, 2021. ISBN 9783030830984. URL https://books.google.com.au/books?id=-N5IEAAAQBAJ.

Peter W. Shor. Algorithms for quantum computation: discrete logarithms and factoring. In *Proceedings 35th Annual Symposium on Foundations of Computer Science*, pages 124–134, 1994. doi: 10.1109/SFCS.1994.365700.

Peter W. Shor. Polynomial-time algorithms for prime factorization and discrete logarithms on a quantum computer. *SIAM Review*, 41(2):303–332, 1999.

Adam Smith, M. S. Kim, Frank Pollmann, and Johannes Knolle. Simulating quantum many-body dynamics on a current digital quantum computer. *npj Quantum Information*, 5(1):106, 2019.

Ranjani G. Sundaram, Himanshu Gupta, and C. R. Ramakrishnan. Distribution of quantum circuits over general quantum networks. *CoRR*, abs/2206.06437, 2022. doi: 10.48550/arXiv.2206.06437.

Rodney Van Meter, W. J. Munro, Kae Nemoto, and Kohei M. Itoh. Arithmetic on a distributed-memory quantum multicomputer. *ACM Journal on Emerging Technologies in Computing Systems*, 3(4), Jan 2008. ISSN 1550-4832. doi: 10.1145/1324177.1324179.

Chonggang Wang and Akbar Rahman. Quantum-enabled 6G wireless networks: opportunities and challenges. *IEEE Wireless Communications*, 29(1):58–69, 2022. doi: 10.1109/MWC.006.00340.

Stephanie Wehner, David Elkouss, and Ronald Hanson. Quantum internet: a vision for the road ahead. *Science*, 362(6412):eaam9288, 2018.

Anocha Yimsiriwattana and Samuel J. Lomonaco Jr. Distributed quantum computing: a distributed Shor algorithm. In Eric Donkor, Andrew R. Pirich, and Howard E. Brandt, editors, *Quantum Information and Computation II*, volume 5436 of *Society of Photo-Optical Instrumentation Engineers (SPIE) Conference Series*, pages 360–372, Aug 2004. doi: 10.1117/12.546504.

Anocha Yimsiriwattana and Samuel J. Lomonaco Jr. Generalized GHZ states and distributed quantum computing. In *Coding Theory and Quantum Computing*, Contemporary Mathematics, pages 131–147. AMS, Providence, RI, 2005. Also available at https://arxiv.org/abs/quant-ph/0402148.

Mingsheng Ying and Yuan Feng. An algebraic language for distributed quantum computing. *IEEE Transactions on Computers*, 58(6):728–743, 2009. doi: 10.1109/TC.2009.13.

2

Preliminaries

This chapter provides background on quantum information and quantum computing to prepare readers for Chapters 3–6. Readers already familiar with this topic might find this chapter unnecessary but going over will at least help in gaining an overview of the conventions and notation we will use throughout this book. Also, the book assumes familiarity with basic linear algebra and probability, but introduces necessary notation and the required mathematics as needed in an attempt to be as self-contained as possible, and also points out references for further reading.

This chapter first introduces the concept of qubits (or quantum bits) and qubit states and then goes on to introduce operations on qubits using quantum gates and quantum circuits, analogous to logic gates and logic circuits operating on bits. We also introduce the notions of entanglement, teleportation, and superdense coding, which are useful concepts in quantum Internet computing and quantum communications.

2.1 Qubit and Qubit States

Classical computing manipulates information represented in a binary manner, using bits, each bit at any one time having one of two possible values, typically denoted by "0" and "1." In contrast, quantum computing manipulates information represented via quantum states, states of qubits, where a *qubit* is a two state or two-level quantum-mechanical system, constituting the basic unit of quantum information, such as a photon with vertical and horizontal polarization, a photon with two paths (discussed below), or electron spin with spin-up and spin-down.

From Distributed Quantum Computing to Quantum Internet Computing: An Introduction,
First Edition. Seng W. Loke.
© 2024 The Institute of Electrical and Electronics Engineers, Inc. Published 2024 by John Wiley & Sons, Inc.

The qubit state can be represented by a linear combination of two basis states $|0\rangle$ and $|1\rangle$,[1] written as follows:

$$\alpha |0\rangle + \beta |1\rangle$$

with *probability amplitudes* $\alpha, \beta \in \mathbb{C}$, where $\mathbb{C}$ is the set of complex numbers,[2] and $|\alpha|^2 + |\beta|^2 = 1$, that is, one can think of the state of the quantum system as a probability distribution over the possible outcomes of measurement done on the system (more on measurement later).

Each basis state can be thought of as a two-dimensional column vector, i.e. the symbols $|0\rangle$ and $|1\rangle$ denote vectors as follows:

$$|0\rangle = \begin{bmatrix} 1 \\ 0 \end{bmatrix} \quad \text{and} \quad |1\rangle = \begin{bmatrix} 0 \\ 1 \end{bmatrix}$$

A qubit state is, hence, represented via a vector in two-dimensional vector space spanned by the basis states. We say that the qubit is in a *superposition* of two states $|0\rangle$ and $|1\rangle$. One can think of the qubit state as representing information encoded in the probability amplitudes.

One can "measure" the state of a qubit, and in doing so, the qubit state changes to a state depending on what basis is used for the measurement. Typically, a qubit is measured in what is known as the *Z basis* or the *computational basis*, in which case, the qubit state changes to either $|0\rangle$ or $|1\rangle$, or more specifically, when measured, the qubit state becomes the state $|0\rangle$ with probability $|\alpha|^2$, or $|1\rangle$ with probability $|\beta|^2$ (which, for simplicity, we briefly say as being due to the "rules of quantum mechanics").

To gain some physical intuition about a qubit, consider an example, where a beam of light is being sent into an apparatus called a *beamsplitter*, which is an optical device that splits a beam of light along two paths, a transmitted beam and a reflected beam, illustrated in Figure 2.1.

A beam of light might contain lots of photons (thinking of a photon as a particle of light), and so, the beamsplitter will cause some of the photons to be reflected and some to be transmitted so that we have the reflected beam of light and the transmitted beam of light; a 50/50 (or *balanced*) beamsplitter would split the incoming beam into two beams, each of lower intensity (compared to the intensity of the incoming beam). However, now consider a "beam of light" with just one photon, then one could consider whether the beamsplitter would cause the photon to travel along the reflected path or the transmitted path, or both. If it was a 50/50 beamsplitter, then it does not split the photon in two, but it will either reflect or transmit the photon with 50% probability for either. One might represent the photon, *after*

---

1 This is the *braket notation* or *Dirac notation* (named after the physicist Paul Dirac), where the symbol $|\psi\rangle$ is a ket and $\langle\psi|$ is a bra, the meanings of which we will explain. We will see that a bra represents the conjugate transpose of the corresponding ket.
2 A complex number is of the form $a + ib$ where $i^2 = -1$, and a and b are real numbers.

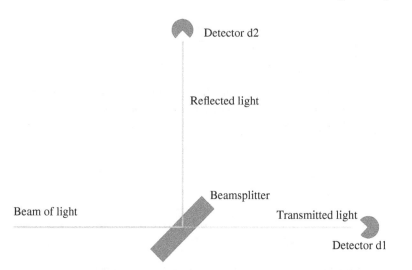

Figure 2.1 A beamsplitter, with detectors d1 and d2 for the transmitted and reflected light, respectively.

the 50/50 beamsplitter, but before measurement with the detectors, as being in the state $\frac{1}{\sqrt{2}}|0\rangle + \frac{1}{\sqrt{2}}|1\rangle$, denoting transmission by $|0\rangle$ and reflection by $|1\rangle$ say, i.e. the photon is in a superposition of the two states until measurement. Then, upon measurement with the detectors, the photon will only be found at one of the detectors, and the probability of $|0\rangle$ or $|1\rangle$ is $(\frac{1}{\sqrt{2}})^2 = \frac{1}{2}$, i.e. detector d1 will detect the photon with probability 0.5, and d2 will detect the photon with probability 0.5.

Upon detection, the state $\frac{1}{\sqrt{2}}|0\rangle + \frac{1}{\sqrt{2}}|1\rangle$ changes to either $|0\rangle$ or to $|1\rangle$, that is, measuring (or observing via the detectors) actually changes the photon's state! In a way, measurement, in this case, is an operation that "transforms" the state of the photon.[3]

Also, suppose we denote the state of the photon before reaching the beamsplitter as $|0\rangle$ (since it is along the same path as the transmitted light) then the beamsplitter changes the state of the photon from $|0\rangle$ to $\frac{1}{\sqrt{2}}|0\rangle + \frac{1}{\sqrt{2}}|1\rangle$. We can represent the operation performed by the beamsplitter via the matrix $H = \frac{1}{\sqrt{2}}\begin{bmatrix} 1 & 1 \\ 1 & -1 \end{bmatrix}$, that is, its

---

3 The idea of measurement usually refers to acquiring information rather than transforming information, and the act of measuring usually should not manipulate the state of the system we are measuring, i.e. we usually measure something that has a certain state (prior to measurement) but simply not known (to us) before measurement – e.g. think of measuring the current temperature. However, "measurement" in quantum theory has the idea of, in a way, manipulating the physical system in some way to obtain an outcome. A starting point for philosophical discussions of measurement in quantum theory is at: https://plato.stanford.edu/entries/qt-issues [last accessed: 19/10/2022].

action on $|0\rangle$ is then:

$$H\,|0\rangle = \frac{1}{\sqrt{2}} \begin{bmatrix} 1 & 1 \\ 1 & -1 \end{bmatrix} \begin{bmatrix} 1 \\ 0 \end{bmatrix} = \frac{1}{\sqrt{2}} \begin{bmatrix} 1 \\ 1 \end{bmatrix} = \frac{1}{\sqrt{2}} \begin{bmatrix} 1 \\ 0 \end{bmatrix} + \frac{1}{\sqrt{2}} \begin{bmatrix} 0 \\ 1 \end{bmatrix} = \frac{1}{\sqrt{2}}\,|0\rangle + \frac{1}{\sqrt{2}}\,|1\rangle$$

The beamsplitter, hence, is also performing some operation on the quantum state, albeit a different kind of operation compared to measurement. Note also that suppose we apply H to a qubit in state $\frac{1}{\sqrt{2}}\,|0\rangle + \frac{1}{\sqrt{2}}\,|1\rangle$, then we have:

$$H\left(\frac{1}{\sqrt{2}}\,|0\rangle + \frac{1}{\sqrt{2}}\,|1\rangle \right) = \frac{1}{\sqrt{2}}(H\,|0\rangle + H\,|1\rangle)$$

$$= \frac{1}{\sqrt{2}} \left(\frac{1}{\sqrt{2}}\,|0\rangle + \frac{1}{\sqrt{2}}\,|1\rangle \right) + \frac{1}{\sqrt{2}} \left(\frac{1}{\sqrt{2}}\,|0\rangle - \frac{1}{\sqrt{2}}\,|1\rangle \right)$$

$$= |0\rangle$$

which is the idea of the $|1\rangle$ component eliminated via "interference," analogous to the idea of interference of light waves resulting in light and dark bands (dark bands where the light has been removed due to interference).[4] This can in fact be done with optical devices in a configuration called the Mach–Zehnder interferometer, as illustrated in Figure 2.2. It is interesting to note that when the photon is in the state $\frac{1}{\sqrt{2}}\,|0\rangle + \frac{1}{\sqrt{2}}\,|1\rangle$, i.e. in a superposition, after the first beamsplitter b1, the photon can be (perhaps naively) thought of as being essentially on "both paths at the same time," even after reflections by the mirrors, until it reaches the second beamsplitter b2 where an interference happens. This idea of "being in two places at the same" time might be rather unintuitive compared to the classical world that we are used to, and when we think of the photon as a particle. Note that $H(H\,|0\rangle) = |0\rangle$ so that H is its own inverse, i.e. $H^{-1} = H$.[5]

Further discussion of the beamsplitter and the interferometer, and their matrix representations, is found in Hughes et al. [2021a].

The upshot of the above is that the qubit state can be manipulated via operations, which in effect, manipulates quantum information (as encoded in the qubit states), and hence, with proper operations on qubits, one achieves *computation, i.e.*

---

4 We may think of the photon as a particle with an associated wave – this illustrates the concept of wave-particle duality in quantum theory where a quantum entity can be described with particle and wave concepts; hence, we might use terminology associated with waves such as "interference" or terminology associated with particles.

5 The discussion on the interferometer provided here might be considered a gross over-simplification, e.g., we are ignoring discussions about phase shifts when light travels from air to glass etc, but we are focusing on the idea that physical devices (e.g., a beamsplitter) can be viewed as "computing something" (in this case, "applying a matrix") (I hope physicists will forgive me!). There are other matrix representations, e.g. with different matrices for the two beamsplitters at https://ocw.mit.edu/courses/8-04-quantum-physics-i-spring-2016/resources/mit8_04s16_lecnotes2 and a matrix for the mirrors in Kauffman and Lomonaco Jr. [2014] [last accessed: 22/10/2022].

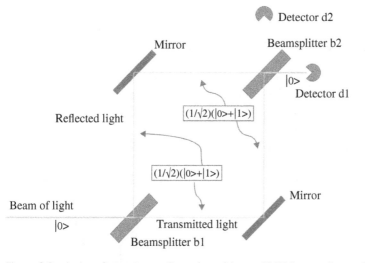

Figure 2.2 An interferometer configuration with two 50/50 beamsplitters. One can represent the first one b1 by the first *H* applied to a photon in state $|0\rangle$ and b2 by the second *H* applied to the photon in the state $\frac{1}{\sqrt{2}}(|0\rangle + |1\rangle)$ to finally obtain back the state $|0\rangle$. Note that the mirrors help to "bring the two components together" as an "input" to the second beamsplitter, and the "output" is the photon in state $|0\rangle$. Note that reflections by the mirrors does not change the state of the photon, and the detectors effectively perform measurement.

the manipulation of information represented by quantum states, and desired computations are achieved with carefully chosen operations on qubits.

One could also have *qutrits*, referring to three state or three-level quantum-mechanical systems, and more generally, *qudits* referring to *d* state or *d* level quantum-mechanical systems.

Also, one can represent the state of a collection of qubits using vectors, e.g. a quantum system comprising a collection of qubits, or a composite system. For example, to represent the state of two qubits, q_1 and q_2, we can use four basis states denoted by $|00\rangle$, $|01\rangle$, $|10\rangle$, and $|11\rangle$, written as follows, using $|\psi\rangle$ to denote the two-qubit state:

$$|\psi\rangle = \alpha |00\rangle + \beta |01\rangle + \gamma |10\rangle + \delta |11\rangle$$

with probability amplitudes α, β, γ, and $\delta \in \mathbb{C}$, and $|\alpha|^2 + |\beta|^2 + |\gamma|^2 + |\delta|^2 = 1$. The four basis states can be represented via four-dimensional vectors:

$$|00\rangle = \begin{bmatrix} 1 \\ 0 \\ 0 \\ 0 \end{bmatrix} \quad \text{and} \quad |01\rangle = \begin{bmatrix} 0 \\ 1 \\ 0 \\ 0 \end{bmatrix}. \quad \text{and} \quad |10\rangle = \begin{bmatrix} 0 \\ 0 \\ 1 \\ 0 \end{bmatrix} \quad \text{and} \quad |11\rangle = \begin{bmatrix} 0 \\ 0 \\ 0 \\ 1 \end{bmatrix}$$

The two-qubit state is, hence, represented as a vector in four-dimensional vector space spanned by the four basis states. In general, the two-qubit state is a superposition of the four states $|00\rangle$, $|01\rangle$, $|10\rangle$, and $|11\rangle$. One might notice that the four basis states are labeled with the binary representation for the numbers 0, 1, 2, and 3, respectively, so that we can write the basis vectors as $|0\rangle$, $|1\rangle$, $|2\rangle$, and $|3\rangle$.

In general, an n-qubit state $|\psi\rangle$ can be represented using a 2^n dimensional vector with basis vectors of the form $|0\rangle, \ldots, |2^n - 1\rangle$, as follows: $|\psi\rangle = \sum_{i=0}^{2^n - 1} \alpha_i |i\rangle$, where $\sum_{i=0}^{2^n - 1} |\alpha_i|^2 = 1$, or using the respective binary string representations: $|\psi\rangle = \sum_{x \in \{0,1\}^n} \alpha_x |x\rangle$, where $\sum_{i=0}^{2^n - 1} |\alpha_x|^2 = 1$. While the state of a qubit is in a vector space,[6] the state of a composite (multiple qubit) system is in a tensor product space.

For multiple qubits, we will use a number of equivalent representations. For example, we have the following equalities:

$$|00\rangle = |0, 0\rangle = |0\rangle |0\rangle = |0\rangle \otimes |0\rangle = \begin{bmatrix} 1 \\ 0 \end{bmatrix} \otimes \begin{bmatrix} 1 \\ 0 \end{bmatrix} = \begin{bmatrix} 1 \begin{bmatrix} 1 \\ 0 \end{bmatrix} \\ 0 \begin{bmatrix} 1 \\ 0 \end{bmatrix} \end{bmatrix} = \begin{bmatrix} 1 \\ 0 \\ 0 \\ 0 \end{bmatrix}$$

where the operator "$\otimes$" denotes the tensor product operator (see **Aside**).

Aside: The *tensor product* of two vector spaces V and W is denoted by $V \otimes W$, which contains elements spanned by vectors of the form $|v\rangle \otimes |w\rangle$, where $|v\rangle \in V$ and $|w\rangle \in W$. The product $|v\rangle \otimes |w\rangle$ is also written as $|v\rangle |w\rangle$, $|v,w\rangle$, or $|vw\rangle$. We recall the following properties of tensor products useful for calculations:

(i) $(|v_1\rangle + |v_2\rangle) \otimes |w\rangle = |v_1\rangle |w\rangle + |v_2\rangle |w\rangle$
(ii) $|v\rangle \otimes (|w_1\rangle + |w_2\rangle) = |v\rangle |w_1\rangle + |v\rangle |w_2\rangle$
(iii) $\alpha(|v\rangle \otimes |w\rangle) = (\alpha |v\rangle) \otimes |w\rangle = |v\rangle \otimes (\alpha |w\rangle)$

Using a matrix representation, given matrices A (m by n) and B, the tensor product $A \otimes B$ is given by

$$A \otimes B = \begin{bmatrix} A_{11}B & A_{12}B & \ldots & A_{1n}B \\ A_{21}B & A_{22}B & \ldots & A_{2n}B \\ \vdots & \vdots & \ldots & \vdots \\ A_{m1}B & A_{m2}B & \ldots & A_{mn}B \end{bmatrix}$$

A tensor product of n copies of a matrix A is denoted by $A^{\otimes n}$. Note that using the fact that $(A \otimes B)(C \otimes D) = AC \otimes BD$, given matrix A and vector v, $A^{\otimes n} v^{\otimes n} = (A \otimes \cdots \otimes A)(v \otimes \cdots \otimes v) = Av \otimes \ldots \otimes Av$.

---

6 More precisely, in quantum theory, the state of a physical system (i.e. qubit) is represented by a vector in a *Hilbert space*, which is a complex vector space with an inner product – see Nielsen and Chuang [2000].

The state of a quantum system (of one or more qubits) $|\psi\rangle$ can also be represented by a matrix of the form $|\psi\rangle\langle\psi|$. In general, suppose there is uncertainty about which state the quantum system is in, then the state of the quantum system can be given by a set of probabilities over the possible states, i.e. a probability p_i for each possible state $|\psi_i\rangle$, called an *ensemble of pure states* $\{p_i, |\psi_i\rangle\}$, which can be represented via a *density operator* or *density matrix* ρ of the form:

$$\rho = \sum_i p_i |\psi_i\rangle\langle\psi_i|$$

where $\sum_i p_i = 1$. The density operator has a number of properties useful in calculations – for example, the trace of ρ, $tr(\rho) = 1$.[7]

When we have a composite system comprising multiple qubits, we can obtain descriptions of the subsystems (some subset of the qubits) of the system using the notion of the *reduced density operator*. For example, let A and B be two physical systems, constituting a composite system, whose state is described by the density operator ρ_{AB}. Then, to describe subsystem A within the composite system, the reduced density operator is as follows:

$$\rho_A = tr_B(\rho_{AB})$$

where the *partial trace over system B* operator, denoted by, tr_B is given by

$$tr_B(|a_1\rangle\langle a_2| \otimes |b_1\rangle\langle b_2|) = |a_1\rangle\langle a_2| \, tr(|b_1\rangle\langle b_2|) = |a_1\rangle\langle a_2| \langle b_2|b_1\rangle$$

for vectors $|a_1\rangle$, $|a_2\rangle$, $|b_1\rangle$, and $|b_2\rangle$, and where tr is the trace operator for matrices.[8] Similarly, we can have an operator $\rho_B = tr_A(\rho_{AB})$ to describe the subsystem B.

2.2 Quantum Gates and Quantum Circuits

We have seen how qubits can be operated on, in the case of a photon and paths using optical devices for operations. Of course, for other physical realizations of the qubit,[9] the operations will done differently, which we will not go into here.

---

7 To see this, consider the state $|\psi\rangle = \begin{bmatrix} \alpha \\ \beta \end{bmatrix}$, then $|\psi_i\rangle\langle\psi_i| = \begin{bmatrix} \alpha \\ \beta \end{bmatrix}\begin{bmatrix} \alpha \\ \beta \end{bmatrix}^\dagger = \begin{bmatrix} \alpha \\ \beta \end{bmatrix}\begin{bmatrix} \bar{\alpha} & \bar{\beta} \end{bmatrix} = \begin{bmatrix} \alpha\bar{\alpha} & \alpha\bar{\beta} \\ \beta\bar{\alpha} & \beta\bar{\beta} \end{bmatrix} =$
$\begin{bmatrix} |\alpha|^2 & \alpha\bar{\beta} \\ \beta\bar{\alpha} & |\beta|^2 \end{bmatrix}$ and since $|\alpha|^2 + |\beta|^2 = 1$, we have $tr(\rho) = 1$, recalling that the trace is the sum of the diagonal elements, and the trace is linear, i.e. $tr(A + B) = tr(A) + tr(B)$, for square matrices A and B.
8 We note that $tr(|b_1\rangle\langle b_2|) = tr(\langle b_2||b_1\rangle) = \langle b_2||b_1\rangle = \langle b_2|b_1\rangle$, using the fact that for matrices P and Q, $tr(PQ) = tr(QP)$ and for a number n, $tr(n) = n$.
9 See a list of current qubit technologies here: https://quantumtech.blog/2022/10/20/quantum-computing-modalities-a-qubit-primer-revisited [last accessed: 22/10/2022]

Analogous to logic gates such as NOT, AND, OR, and NAND gates in digital logic design that operate on bits, operations on qubits have been identified as *quantum gates*. We present the gates in this section via their matrix representations, without looking at how they might be physically realized. Just as logic gates can be composed to form logic circuits, we will see how quantum gates can be composed to form *quantum circuits*; we also introduce symbols we will use for different quantum gates in quantum circuits.

2.2.1 Single Qubit Gates

We have seen the operation we called H which has been called the *Hadamard* gate.[10] We consider other gates, focusing on those used in this book below.

The quantum NOT gate X can be represented as follows:

$$X = \begin{bmatrix} 0 & 1 \\ 1 & 0 \end{bmatrix}$$

which has the following effect on a quantum state $\alpha |0\rangle + \beta |1\rangle$:

$$X(\alpha |0\rangle + \beta |1\rangle) = \alpha(X |0\rangle) + \beta(X |1\rangle) = \alpha |1\rangle + \beta |0\rangle$$

since $X |0\rangle = |1\rangle$ and $X |1\rangle = |0\rangle$ which can be verified via matrix multiplication.

The quantum gate Z can be represented as follows:

$$Z = \begin{bmatrix} 1 & 0 \\ 0 & -1 \end{bmatrix}$$

which has the following effect on a quantum state $\alpha |0\rangle + \beta |1\rangle$:

$$Z(\alpha |0\rangle + \beta |1\rangle) = \alpha(Z |0\rangle) + \beta(Z |1\rangle) = \alpha |0\rangle - \beta |1\rangle$$

since $Z |0\rangle = |0\rangle$ and $Z |1\rangle = -|1\rangle$ which can be verified via matrix multiplication.

The quantum gate Y can be represented as follows:

$$Y = \begin{bmatrix} 0 & -i \\ i & 0 \end{bmatrix}$$

which has the following effect on a quantum state $\alpha |0\rangle + \beta |1\rangle$:

$$Y(\alpha |0\rangle + \beta |1\rangle) = \alpha(Y |0\rangle) + \beta(Y |1\rangle) = \alpha i |1\rangle - \beta i |0\rangle$$

since $Y |0\rangle = i |1\rangle$ and $Y |1\rangle = -i |0\rangle$ which can be verified via matrix multiplication.

Note that the X, Y, and Z matrices are famously called the *Pauli matrices*, named after the physicist Wolfgang Pauli.

---

10 The matrix is named after the French mathematician Jacques Hadamard.

When we apply the H gate to $\alpha \left|0\right\rangle + \beta \left|1\right\rangle$, we have

$$H(\alpha \left|0\right\rangle + \beta \left|1\right\rangle) = \alpha (H \left|0\right\rangle) + \beta (H \left|1\right\rangle) = \frac{\alpha + \beta}{\sqrt{2}} \left|0\right\rangle + \frac{\alpha - \beta}{\sqrt{2}} \left|1\right\rangle$$

since $H \left|0\right\rangle = \frac{1}{\sqrt{2}} \left|0\right\rangle + \frac{1}{\sqrt{2}} \left|1\right\rangle$ and $H \left|1\right\rangle = \frac{1}{\sqrt{2}} \left|0\right\rangle - \frac{1}{\sqrt{2}} \left|1\right\rangle$, which can be verified via matrix multiplication. Note that $\left|\frac{\alpha+\beta}{\sqrt{2}}\right|^2 + \left|\frac{\alpha-\beta}{\sqrt{2}}\right|^2 = 1$, since $\left|\alpha\right|^2 + \left|\beta\right|^2 = 1$.[11]

We will introduce other gates, e.g. what we call the R gate and P gate, later in the book as needed. In general, a single qubit quantum gate U is typically represented as a 2×2 matrix with complex numbers as entries and is *unitary*, that is, its inverse is the same as its conjugate transpose: $U^{-1} = U^\dagger$ or $UU^{-1} = UU^\dagger = U^\dagger U = I$, where $U^\dagger$ denotes the *conjugate transpose* (or *Hermitian transpose*) of U, which is formed by transposing U and taking the complex conjugate[12] of all its entries, i.e. $U^\dagger = \overline{U^T}$. For unitary U, we have $UU^\dagger = U^\dagger U = I$, where I denotes the identity matrix. Also, a square matrix is *Hermitian* if it is its own conjugate transpose, i.e. $U = U^\dagger$. Note that the gates we have seen X, Z, Y, and H are all Hermitian and unitary, i.e. $XX = ZZ = YY = HH = I$ and are their own inverses.

It is also noted that for a unitary matrix, its action on a state vector $\mathbf{s} = \alpha \left|0\right\rangle + \beta \left|1\right\rangle = \begin{pmatrix} \alpha \\ \beta \end{pmatrix}$ preserves the norm, that is, $||U\mathbf{s}|| = ||\mathbf{s}||$, where $||.||$ denotes the norm of a vector, i.e. for a vector $\mathbf{x}$, $||\mathbf{x}|| = \sqrt{\langle \mathbf{x}, \mathbf{x} \rangle} = \sqrt{\mathbf{x}^\dagger \mathbf{x}}$, where $\langle \cdot, \cdot \rangle$ denotes the inner product.[13] Hence, unitary operators help to preserve the property that the squares of the probability amplitudes sum to one: suppose U and $\mathbf{s}$ are such that $||\mathbf{s}||^2 = \begin{pmatrix} \alpha \\ \beta \end{pmatrix}^\dagger \begin{pmatrix} \alpha \\ \beta \end{pmatrix} = \alpha\bar{\alpha} + \beta\bar{\beta} = \left|\alpha\right|^2 + \left|\beta\right|^2 = 1$, $||U\mathbf{s}|| = ||\mathbf{s}||$ and $U\mathbf{s} = \begin{pmatrix} \alpha' \\ \beta' \end{pmatrix}$, then $||U\mathbf{s}|| = \left|\alpha'\right|^2 + \left|\beta'\right|^2 = ||\mathbf{s}|| = 1$. It therefore makes sense that quantum gates mapping one valid quantum state to another valid quantum state would need to be unitary operators.

An intuitive interpretation of a quantum gate is as manipulation of the state vector $\left|\psi\right\rangle = \alpha \left|0\right\rangle + \beta \left|1\right\rangle$, as we have seen, within some state space, or some coordinate system. But consider the *Bloch sphere*, which is a geometrical representation of the state space of a qubit, where the two poles of the sphere correspond to the two basis vectors $\left|0\right\rangle$ and $\left|1\right\rangle$, and then an arrow on the Bloch sphere (starting from the center of the sphere) represents a state of the qubit. Many books have explained

---

11 One can show this by setting $\alpha = a_1 + ia_2$ and $\beta = b_1 + ib_2$ and noting that $\left|\alpha\right|^2 = a_1^2 + a_2^2$ and $\left|\beta\right|^2 = b_1^2 + b_2^2$, $\left|\alpha + \beta\right|^2 = \left|a_1 + ia_2 + b_1 + ib_2\right|^2 = \left|a_1 + b_1 + i(a_2 + b_2)\right|^2 = (a_1 + b_1)^2 + (a_2 + b_2)^2$ and $\left|\alpha - \beta\right|^2 = \left|(a_1 + ia_2) - (b_1 + ib_2)\right|^2 = \left|(a_1 - b_1) + i(a_2 - b_2)\right|^2 = (a_1 - b_1)^2 + (a_2 - b_2)^2$, so that $\left|\frac{\alpha+\beta}{\sqrt{2}}\right|^2 + \left|\frac{\alpha-\beta}{\sqrt{2}}\right|^2 = \frac{1}{2}(\left|\alpha + \beta\right|^2 + \left|\alpha - \beta\right|^2) = \frac{1}{2}(2a_1^2 + 2b_1^2 + 2a_2^2 + 2b_2^2) = a_1^2 + a_2^2 + b_1^2 + b_2^2 = \left|\alpha\right|^2 + \left|\beta\right|^2 = 1$.

12 The complex conjugate of a complex number $c = a_1 + ia_2$ is $\bar{c} = a_1 - ia_2$, which is denoted with an overline $\bar{c}$. We will also use the notation c^* for complex conjugate, when convenient.

13 To see this, note that $||\mathbf{s}||^2 = \langle \mathbf{s}, \mathbf{s} \rangle = \mathbf{s}^\dagger \mathbf{s} = \mathbf{s}^\dagger I \mathbf{s} = \mathbf{s}^\dagger U^\dagger U \mathbf{s} = (U\mathbf{s})^\dagger U\mathbf{s} = \langle U\mathbf{s}, U\mathbf{s} \rangle = ||U\mathbf{s}||^2$, where we have used the identity: for matrices A and B, $A^\dagger B^\dagger = (BA)^\dagger$.

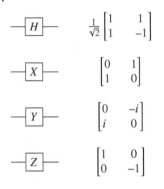

Figure 2.3 Symbols and matrices for single qubit gates H, X, Y, and Z.

the Bloch sphere and so we will not repeat this here, but refer readers to Hughes et al. [2021b]. With such a geometrical representation, quantum gate operations can be thought of as rotations of the arrow around the three orthogonal axes (x, y, and z) running through the center of the sphere.[14]

Diagrammatically, the symbols for the quantum gates H, X, Y, and Z are shown in Figure 2.3.

Note that when quantum state is represented via a *density operator*, say ρ of the form: $\rho = \sum_i p_i \, |\psi_i\rangle \langle\psi_i|$, where $\sum_i p_i = 1$, then evolution of the quantum system to a new state due to a unitary operator U is then given by

$$U\rho U^\dagger = U \left(\sum_i p_i \, |\psi_i\rangle \langle\psi_i| \right) U^\dagger = \sum_i p_i U \, |\psi_i\rangle \langle\psi_i| \, U^\dagger$$

2.2.2 Measurement Operators

We saw earlier the idea of measurement which effectively "transforms" a state $|\psi\rangle = \alpha \, |0\rangle + \beta \, |1\rangle$ into either $|0\rangle$ with probability $|\alpha|^2$ or $|1\rangle$ with probability $|\beta|^2$.

Letting $\langle 0| = |0\rangle^\dagger$ and $\langle 1| = |1\rangle^\dagger$, one can represent measurement using operators often denoted in the following form:

$$|0\rangle \langle 0| = \begin{bmatrix} 1 \\ 0 \end{bmatrix} \begin{bmatrix} 1 & 0 \end{bmatrix} = \begin{bmatrix} 1 & 0 \\ 0 & 0 \end{bmatrix} \quad \text{and} \quad |1\rangle \langle 1| = \begin{bmatrix} 0 \\ 1 \end{bmatrix} \begin{bmatrix} 0 & 1 \end{bmatrix} = \begin{bmatrix} 0 & 0 \\ 0 & 1 \end{bmatrix}$$

Note that $|0\rangle \langle 0| + |1\rangle \langle 1| = I$. Then, since $(|0\rangle \langle 0|)(\alpha \, |0\rangle + \beta \, |1\rangle) = \alpha \, |0\rangle \langle 0| \, |0\rangle + \beta \, |0\rangle \langle 0| \, |1\rangle = \alpha \, |0\rangle$ and

---

14 In fact, any unitary operation U on a single qubit can be expressed as a composition of rotation operators about the three axes: $U = e^{i\alpha} R_z(\beta) R_y(\gamma) R_z(\delta)$, for some real numbers $\alpha, \beta, \gamma,$ and δ, which specify the rotations in the rotation operators $R_z(\beta)$, $R_y(\gamma)$, and $R_z(\delta)$ [Nielsen and Chuang, 2000].

$$(|0\rangle \langle 0| (\alpha |0\rangle + \beta |1\rangle))^\dagger = (\alpha |0\rangle + \beta |1\rangle)^\dagger (|0\rangle \langle 0|)^\dagger$$
$$= (\alpha |0\rangle + \beta |1\rangle)^\dagger \langle 0|^\dagger |0\rangle^\dagger = (\overline{\alpha} \langle 0| + \overline{\beta} \langle 1|) |0\rangle \langle 0|$$
$$= \overline{\alpha} \langle 0| |0\rangle \langle 0| + \overline{\beta} \langle 1| |0\rangle \langle 0| = \overline{\alpha} \langle 0|$$

the probability of an outcome resulting in $|0\rangle$ can be written as

$$prob(|0\rangle) = |\alpha|^2 = \overline{\alpha}\alpha = \overline{\alpha}\alpha \langle 0| |0\rangle = \overline{\alpha} \langle 0| \alpha |0\rangle$$
$$= (\alpha |0\rangle + \beta |1\rangle)^\dagger (|0\rangle \langle 0|)^\dagger (|0\rangle \langle 0|)(\alpha |0\rangle + \beta |1\rangle)$$
$$= \langle \psi | M_0{}^\dagger M_0 |\psi\rangle$$

where the last line is obtained by setting $M_0 = |0\rangle \langle 0|$. Similarly, let $M_1 = |1\rangle \langle 1|$, and a similar calculation will yield: $prob(|1\rangle) = \langle \psi | M_1{}^\dagger M_1 |\psi\rangle$.

2.2.2.1 Measurement Postulate (General Measurement)

More generally, a postulate in quantum theory is that quantum measurements can be represented by a set of *measurement operators* $\{M_m\}$, each acting on the states of the quantum system being measured, and the index m refers to a measurement outcome that may occur with respect to operator M_m, and that the probability of outcome m is given by

$$p(m) = \langle \psi | M_m{}^\dagger M_m |\psi\rangle$$

and the state after measurement is given by

$$\frac{M_m |\psi\rangle}{\sqrt{\langle \psi | M_m{}^\dagger M_m |\psi\rangle}} = \frac{M_m |\psi\rangle}{\sqrt{p(m)}}$$

and the operators satisfy the following *completeness* condition: $\sum_m M_m{}^\dagger M_m = I$.

Coming back to our example, after M_0, we have the state:

$$\frac{M_0 |\psi\rangle}{\sqrt{\langle \psi | M_0{}^\dagger M_0 |\psi\rangle}} = \frac{M_0(\alpha |0\rangle + \beta |1\rangle)}{\sqrt{|\alpha|^2}} = \frac{\alpha |0\rangle}{|\alpha|}$$

and after M_1, we would have the state: $\frac{\beta |1\rangle}{|\beta|}$. Since $|\frac{\alpha}{|\alpha|}| = |\frac{\beta}{|\beta|}| = 1$, we may ignore the multipliers, obtaining the state $|0\rangle$ or $|1\rangle$ as we saw earlier.

2.2.2.2 POVM

Note that if we define a set of positive operators $\{E_m\}$ say by $E_m = M_m{}^\dagger M_m$, these operators are called *POVM (Positive Operator-Valued Measure) elements* and $\{E_m\}$ is called a POVM.

2.2.2.3 Projective Measurements

Suppose we have a set of measurement operators $\{P_m\}$, where there is a Hermitian matrix M (representing an *observable*, a quantity that can be measured, in Physics) as follows:

$$M = \sum_m m P_m$$

which shows how this set of measurement operators $\{P_m\}$ are related, and the probability of outcome m is given by $p(m) = \langle \psi | P_m | \psi \rangle$, and and the state after measurement is given by $\frac{P_m | \psi \rangle}{\sqrt{p(m)}}$, then M describes a *projective measurement* with measurement operators $\{P_m\}$ which are projectors onto the eigenspace of M with eigenvalue m (e.g. denoting the corresponding eigenvector of M by $|m\rangle$, i.e. $M|m\rangle = m|m\rangle$, then we may write P_m as $|m\rangle\langle m|$).

Also, suppose we have a set of measurement operators $\{M_m\}$ according to the postulate above and that the operators are orthogonal projectors, i.e. $M_m M_{m'} = \delta_{m,m'} M_m$, where $\delta_{m,m'} = 1$ if $m = m'$ and $\delta_{m,m'} = 0$ if $m \neq m'$. Then, note that $p(m) = \langle \psi | M_m^\dagger M_m | \psi \rangle = \langle \psi | M_m | \psi \rangle$, since $M_m^\dagger M_m = M_m M_m = M_m$ (noting that $M_m^\dagger = M_m$, since M_m is Hermitian). Hence, projective measurements form a special class of measurements.

An example of a projective measurement is the measurement in the *computational basis* using measurement operators $\{M_0 = |0\rangle\langle 0|, M_1 = |1\rangle\langle 1|\}$ or measurement with respect to the observable $Z = |0\rangle\langle 0| - |1\rangle\langle 1| = M_0 - M_1$, where acting on $\alpha|0\rangle + \beta|1\rangle$, we observe that the probability of getting an outcome corresponding to eigenvalue 1 is $p(1) = \bar{a}a = |\alpha|^2$ with resulting state $|0\rangle$, and the probability of getting an outcome corresponding to eigenvalue -1 is $p(-1) = |\beta|^2$ with resulting state $|1\rangle$.

Another example is the measurement in the *Hadamard basis*, i.e. with respect to the measurement operators $\{|+\rangle\langle +|, |-\rangle\langle -|\}$, where $|+\rangle = (|0\rangle + |1\rangle)/\sqrt{2}$ and $|-\rangle = (|0\rangle - |1\rangle)/\sqrt{2}$. The observable is the matrix $X = |+\rangle\langle +| - |-\rangle\langle -|$. What are the probabilities of outcomes if we measure a qubit in the state $\alpha|0\rangle + \beta|1\rangle$? We can rewrite the state as follows: $\alpha|0\rangle + \beta|1\rangle = \frac{\alpha+\beta}{\sqrt{2}}|+\rangle + \frac{\alpha-\beta}{\sqrt{2}}|-\rangle$, and we observe that the probability of getting an outcome corresponding to eigenvalue 1 is $p(1) = |\frac{\alpha+\beta}{\sqrt{2}}|^2$ with resulting state $|+\rangle$, and the probability of getting an outcome corresponding to eigenvalue -1 is $p(-1) = |\frac{\alpha-\beta}{\sqrt{2}}|^2$ with resulting state $|-\rangle$. Note that suppose our instruments can only measure in the computational basis, then, what one can do is to first apply the (Hadamard) H operator to the state as follows:

$$H(\alpha|0\rangle + \beta|1\rangle) = \alpha H|0\rangle + \beta H|1\rangle = \alpha|+\rangle + \beta|-\rangle = \frac{\alpha+\beta}{\sqrt{2}}|0\rangle + \frac{\alpha-\beta}{\sqrt{2}}|1\rangle$$

and then, do measurements with respect to the computational basis. Note that now the probability of getting an outcome corresponding to eigenvalue 1 is

$p(1) = |\frac{\alpha+\beta}{\sqrt{2}}|^2$ with resulting state $|0\rangle$, and the probability of getting an outcome corresponding to eigenvalue -1 is $p(-1) = |\frac{\alpha-\beta}{\sqrt{2}}|^2$ with resulting state $|1\rangle$, i.e. the same probabilities as when we were measuring in the Hadamard basis! This means that measuring with respect to the Hadamard basis is effectively equivalent (except for the resulting states) to first applying H to the state and then measuring with respect to the computational basis.

Hence, a key observation from the last two paragraphs is that the same state can be measured with respect to different bases, with potentially different outcome states (corresponding to the basis states) and probabilities of outcomes. A more extensive discussion of measurement operators is in Nielsen and Chuang [2000]. The symbol we will use for measurement is as shown in Figure 2.4.

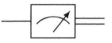

Figure 2.4 Symbol for a quantum measurement operation, where the input is a quantum state (possibly a superposition) but the resulting state is a basis state corresponding to the measurement performed, and a value corresponding to the measurement outcome.

2.2.3 Multiple Qubit Gates

Just like we have logic gates that take more than one operand, we have quantum gates that can take multiple qubits as input. One example is the controlled-NOT (or *CNOT*) gate which takes two qubits, one called a *control qubit* and another called a *target qubit*, and depending on the state of the control qubit, determines whether to apply an X gate to the target qubit. It does the following mapping, where reading from left to right, the first qubit is the control qubit and the second qubit is the target qubit: $|00\rangle \rightarrow |00\rangle, |01\rangle \rightarrow |01\rangle, |10\rangle \rightarrow |11\rangle, |11\rangle \rightarrow |10\rangle$, or in general: $|c\rangle |t\rangle \xrightarrow{CNOT} |c\rangle |t \oplus c\rangle$, where "$\oplus$" denotes the exclusive OR operator (or addition modulo 2). The symbol for a CNOT gate is given in Figure 2.5.

More generally, we can also have a *controlled-U gate* of the form: $|c\rangle |t\rangle \xrightarrow{controlled-U} |c\rangle U |t\rangle$. The *CNOT* gate is basically a controlled-X gate. The symbol for a controlled-U gate is given in Figure 2.5.

2.2.4 Quantum Circuits

We have seen some examples of symbols for quantum gates. When these gates are composed and the gates are acting on one or more qubits, we can construct

Figure 2.5 Symbols for the CNOT gate and controlled-U gate.

quantum circuits comprising a composition of quantum gates, analogous to digital circuits comprising a composition of logic gates. Each line in a quantum circuit is a *wire* representing a qubit and the passage of time is typically from left to right, and the order of computations as shown in a circuit is read from left to right (note that this is in reverse order to a sequence of matrices representing the operations) as we illustrate below. As an example, consider the following two-qubit circuit, shown in Figure 2.6.

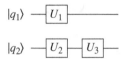

Figure 2.6 An example circuit with two wires for two qubits.

Equivalently, supposing the state of the two-qubit system can be written as $|q_1\rangle \otimes |q_2\rangle$, the operations represented by the circuit can be written in matrix form, as follows:

$$(I \otimes U_3)(U_1 \otimes U_2)(|q1\rangle \otimes |q2\rangle)$$

which is the product of the matrices $I \otimes U_3$, and $U_1 \otimes U_2$ in the order shown above – we used the identity matrix I since there is no operation on that qubit at that stage. Using the property about the product of tensor products, i.e. $(A \otimes B)(C \otimes D) = AC \otimes BD$, we rewrite the above as $(I \otimes U_3)(U_1 \otimes U_2)(|q1\rangle \otimes |q2\rangle) = U_1 |q1\rangle \otimes U_3 U_2 |q2\rangle$. But as we shall see when we talk about entanglement that not all two-qubit states can be written as a product of two vectors of the form $|v\rangle \otimes |w\rangle = |vw\rangle$, but more generally, a two-qubit state $|\psi\rangle$ is a linear combination of such products, as we have seen, e.g. with respect to the computational basis, $|\psi\rangle = \alpha |00\rangle + \beta |01\rangle + \gamma |10\rangle + \delta |11\rangle$; a relabeling is shown in Figure 2.7a. In fact, for each component of the superposition, one could imagine a corresponding circuit as illustrated in Figure 2.7b. Hence, one might consider Figure 2.7a as an "abbreviation" for Figure 2.7b.

That is, in this case, we have

$$(U_1 \otimes U_3 U_2) |\psi\rangle = (U_1 \otimes U_3 U_2)(\alpha |00\rangle + \beta |01\rangle + \gamma |10\rangle + \delta |11\rangle)$$
$$= \alpha(U_1 \otimes U_3 U_2) |00\rangle + \beta(U_1 \otimes U_3 U_2) |01\rangle$$
$$+ \gamma(U_1 \otimes U_3 U_2) |10\rangle + \delta(U_1 \otimes U_3 U_2) |11\rangle$$

with probabilities corresponding to the components in Figure 2.7b.

We consider another example of a circuit; the following three qubit circuit, shown in Figure 2.8. We can see how the single and double qubit gates we mentioned can be combined to form quantum circuits, representing operations over a set of qubits.

Note also that the operations represented by an n-qubit quantum circuit corresponds to an n-qubit unitary operation; the tensor product of unitary matrices is

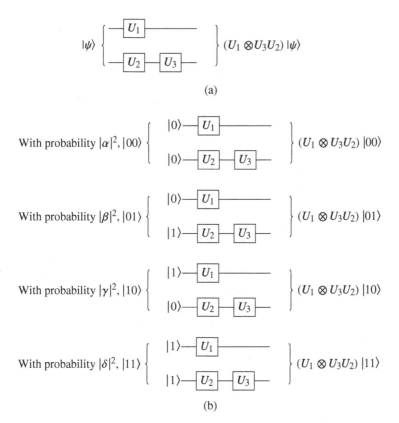

(a)

(b)

Figure 2.7 An example of a two-qubit circuit. (a) An example circuit with two wires for two qubits with an input two-qubit state and label for the new state after the operations of the circuit. (b) Illustration of four circuits with two wires, one for each possible component in a superposition $\alpha |00\rangle + \beta |01\rangle + \gamma |10\rangle + \delta |11\rangle$ with the new state after the operations of the circuit, for each possible component.

also a unitary matrix,[15] and the product of two unitary matrices is also a unitary matrix.[16]

It is interesting to observe the following circuit which is used later in the book, shown in Figure 2.9, where we have $|\psi\rangle = \alpha |0\rangle + \beta |1\rangle$, which computes the CNOT on $|\psi\rangle |0\rangle = (\alpha |0\rangle + \beta |1\rangle) |0\rangle = \alpha |00\rangle + \beta |10\rangle$, with the resulting state $\alpha |00\rangle + \beta |11\rangle$. Note that the circuit does not, in general, actually compute a

---

15 We can see this since $(A \otimes B)^{\dagger} = A^{\dagger} \otimes B^{\dagger}$, we have $(A \otimes B)(A \otimes B)^{\dagger} = (A \otimes B)(A^{\dagger} \otimes B^{\dagger}) = AA^{\dagger} \otimes BB^{\dagger} = I_{dim(A)} \otimes I_{dim(B)} = I_{dim(A) \times dim(B)}$, and similar calculations for $(A \otimes B)^{\dagger}(A \otimes B)$.
16 To see this, since $(AB)^{\dagger} = B^{\dagger}A^{\dagger}$, we have $AB(AB)^{\dagger} = ABB^{\dagger}A^{\dagger} = AIA^{\dagger} = AA^{\dagger} = I$.

copy of $|\psi\rangle$ since

$$|\psi\rangle\,|\psi\rangle = \alpha^2\,|00\rangle + \alpha\beta\,|01\rangle + \beta\alpha\,|10\rangle + \beta^2\,|11\rangle$$

which is equal to $\alpha\,|00\rangle + \beta\,|11\rangle$ when $\alpha\beta = 0$, $\alpha^2 = \alpha$, and $\beta^2 = \beta$, e.g. when $\alpha = 1$ and $\beta = 0$, or when $\alpha = 0$ and $\beta = 1$.

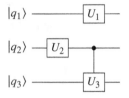

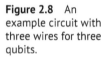

Figure 2.8 An example circuit with three wires for three qubits.

In fact, there is no unitary operation that will, for any quantum state $|\psi\rangle$, take $|\psi\rangle\,|0\rangle$ and yield $|\psi\rangle\,|\psi\rangle$, a fact called the *no-cloning theorem* – see a proof in Nielsen and Chuang [2000]. This theorem has consequences for robust computations and quantum communications since making copies of information is a typical way to deal with lost or corrupted information, at least with classical information. The area of study called *quantum error correction* looks at ways to deal with errors in quantum information. If obtaining $|\psi\rangle\,|\psi\rangle$ was possible, then if, say, the first qubit was lost (or somehow "measured"), we still retain the information in the second qubit $|\psi\rangle$ (the notation $|\psi\rangle\,|\psi\rangle = |\psi\rangle \otimes |\psi\rangle$ essentially suggests the two qubits are "separate"), but if we have $\alpha\,|00\rangle + \beta\,|11\rangle$, then losing the first qubit (effectively, somehow "measured out"), the state "collapses" to $|00\rangle$ or $|11\rangle$ and the first qubit lost, then we either have $|0\rangle$ or $|1\rangle$ remaining, basically losing all the information encoded in the probability amplitudes α and β. We will see many other quantum circuits in this book for different computations.

$$|\psi\rangle = \alpha\,|0\rangle + \beta\,|1\rangle$$
$$|0\rangle$$
$$\Big\} \; \alpha\,|00\rangle + \beta\,|11\rangle \; (= |\psi\rangle\,|\psi\rangle?)$$

Figure 2.9 An example circuit which *looks like* a copying operation.

2.2.5 Universal Quantum Computer and Gate Sets

By composing gates in a quantum circuit, complex manipulations can be performed on qubits. The gates composed can be drawn from a small finite set, and with the right set of gates, any unitary operation can be approximated (to arbitrary accuracy) by composing gates (drawn from the set) together in a quantum circuit.

The Solovay–Kitaev theorem states that, given a distance ϵ (using a distance measure representing the difference between two operations, which we will not go into detail here), one can approximate any given unitary operation on n qubits using $O(n^2 4^n \log^c(n^2 4^n / \epsilon))$ gates [Nielsen and Chuang, 2000]. This is analogous to how a small set of logic gates (e.g. $\{AND, OR, NOT\}$) can be considered universal in that one can construct any digital circuit (i.e. implementing any Boolean function of a particular number of bits) just by using gates from that set (of course, using the same gate multiple times is allowed). The NAND gate on its own is universal. There could be different possible universal gate sets.

One such universal gate set for quantum computations comprises the Hadamard, phase, CNOT, and $\pi/8$ gates; we will look at the phase and $\pi/8$ gates later in the book. With such a universal gate set, in principle, if one can implement each gate in the gate set with some hardware, and execute quantum circuits involving gates from that gate set, one then has a universal quantum computer. It is interesting to note that any two-qubit (generic) gate is adequate for universal quantum computation, i.e. we simply keep using this generic gate between potentially different pairs of qubits.[17] There are many possible universal gate sets.[18]

2.3 Entanglement

We referred to the notion of entanglement previously, and also saw the state $\alpha\,|00\rangle + \beta\,|11\rangle$. Consider the popular *entangled* state, denoted by $|\Phi^+\rangle$, which is a *Bell state* (or a *Bell pair*) (named after John Bell we mentioned in Chapter 1): $|\Phi^+\rangle = \frac{1}{\sqrt{2}}\,|00\rangle + \frac{1}{\sqrt{2}}\,|11\rangle$. When the two qubits are in the above state, we say that the two qubits have been entangled, and has the following sense. Suppose two photons interacted so that their joint state is represented by the above superposition. We can write this superposition with subscripts for the two qubits A and B as follows: $|\Phi^+\rangle_{AB} = \frac{1}{\sqrt{2}}\,|0\rangle_A|0\rangle_B + \frac{1}{\sqrt{2}}\,|1\rangle_A|1\rangle_B$. Then, suppose the two photons fly very far apart, and say, are now billions of light years apart. Then, when A is measured, note that the two-qubit state changes to either $|0\rangle_A|0\rangle_B$ with probability $(1/\sqrt{2})^2 = 1/2$ or to $|1\rangle_A|1\rangle_B$ with probability $1/2$. In other words, when measured, the states of the two qubits are either both $|0\rangle$ or both $|1\rangle$, i.e. they are correlated even though they would not have been able to communicate in time (with respect to the time of observed measurements, being so far apart).[19]

---

17 For definition of "generic" gate and more details on universal quantum gate sets, see Preskill's Caltech notes at http://theory.caltech.edu/~preskill/ph229/notes/chap6.pdf [last accessed: 26/10/2022]

18 A discussion on universal gate sets and some basic gate sets are at: https://quantumcomput ing.stackexchange.com/questions/17115/basic-gates-sets/17126 [last accessed: 27/2/2023]

19 We mentioned this in Chapter 1 as what Einstein spoke about as "spooky action at a distance" as the measurement of one seems to have "acted" on the other even over vast distances, but thinking about this, alternatively, it seems that the two qubits, even though far apart, are behaving like "one entity" with shared state $|1\rangle_A|1\rangle_B$ or $|0\rangle_A|0\rangle_B$ upon measurement, which relates to the idea of *non-locality* in quantum physics, since the "two parts" are so far apart yet can be correlated.

We say that the two qubits from A and B are entangled, or there is entanglement between the two qubits at A and B; this is also referred to as one *ebit*, i.e. having x ebits corresponds to having x entangled pairs.

Can we write the state $|\Phi^+\rangle$ as a tensor product of two vectors? Suppose we can and the two vectors are $\alpha_1 |0\rangle + \beta_1 |1\rangle$ and $\alpha_2 |0\rangle + \beta_2 |1\rangle$, which have tensor product of the form: $(\alpha_1 |0\rangle + \beta_1 |1\rangle) \otimes (\alpha_2 |0\rangle + \beta_2 |1\rangle) = \alpha_1\alpha_2 |00\rangle + \alpha_1\beta_2 |01\rangle + \beta_1\alpha_2 |10\rangle + \beta_1\beta_2 |11\rangle$, but for

$$|\Phi^+\rangle = \frac{1}{\sqrt{2}} |00\rangle + \frac{1}{\sqrt{2}} |11\rangle = \alpha_1\alpha_2 |00\rangle + \alpha_1\beta_2 |01\rangle + \beta_1\alpha_2 |10\rangle + \beta_1\beta_2 |11\rangle$$

we would need to have: $\frac{1}{\sqrt{2}} = \alpha_1\alpha_2, 0 = \alpha_1\beta_2, 0 = \beta_1\alpha_2$, and $\frac{1}{\sqrt{2}} = \beta_1\beta_2$, that is, from the second equation, $\alpha_1 = 0$ or $\beta_2 = 0$, but the first equation cannot be true if $\alpha_1 = 0$ and the fourth equation cannot be true if $\beta_2 = 0$, and so, we have a contradiction. This means that $|\Phi^+\rangle$ cannot be represented as a tensor product of two vectors. In general, a (two-qubit) entangled state is one which cannot be written as a tensor product of two vectors; this idea can be generalized to an n-qubit entangled state. Note that $|\Phi^+\rangle$ cannot be written as *a* "tensor product of two vectors" but can be written as a linear combination of *two* "tensor product of two vectors," namely, a linear combination of $|0\rangle \otimes |0\rangle$ and $|1\rangle \otimes |1\rangle$. More generally, via a similar argument, note that the state $\alpha |00\rangle + \beta |11\rangle$ is entangled for $\alpha \neq 0$ and $\beta \neq 0$. A state in this form is called a *cat-like state*,[20] which will be particularly useful in Chapter 4. The density operator corresponding to $|\Phi^+\rangle$ is

$$|\Phi^+\rangle\langle\Phi^+| = \left(\frac{1}{\sqrt{2}} |00\rangle + \frac{1}{\sqrt{2}} |11\rangle\right)\left(\frac{1}{\sqrt{2}} \langle 00| + \frac{1}{\sqrt{2}} \langle 11|\right) = \frac{1}{2}\begin{bmatrix} 1 & 0 & 0 & 1 \\ 0 & 0 & 0 & 0 \\ 0 & 0 & 0 & 0 \\ 1 & 0 & 0 & 1 \end{bmatrix}$$

In terms of computation, such an entangled state can be created via a circuit shown in Figure 2.10. using a Hadamard gate and a CNOT gate acting on two qubits both in state $|0\rangle$.

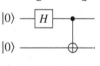

Figure 2.10 A circuit to generate the entangled state $|\Phi^+\rangle = \frac{1}{\sqrt{2}}(|00\rangle + |11\rangle)$.

---

20 See examples of experimental realization of cat states and a concise description from Leibfried et al. [2005]: "Among such states, the so-called cat states, named after Schrödinger's cat, are of particular interest. Cat states are equal superpositions of two maximally different states... and play a distinguished role in quantum information science. For three ion-qubits, they are also called Greenberger–Horne–Zeilinger (GHZ) states and provide a particularly clear demonstration of quantum non-locality."

This can be seen via calculations:

$$CNOT(H \otimes I)(|0\rangle \otimes |0\rangle)$$

$$= CNOT(H|0\rangle \otimes |0\rangle) = CNOT(\frac{1}{\sqrt{2}}(|0\rangle + |1\rangle) \otimes |0\rangle)$$

$$= CNOT(\frac{1}{\sqrt{2}}(|0\rangle |0\rangle + |1\rangle |0\rangle)) = \frac{1}{\sqrt{2}}(CNOT |0\rangle |0\rangle + CNOT |1\rangle |0\rangle)$$

$$= \frac{1}{\sqrt{2}}(|00\rangle + |11\rangle)$$

which is the entangled state we wanted to generate. The four Bell states (also called Einstein–Podolsky–Rosen (EPR) pairs[21]) are $|\Phi^+\rangle = \frac{1}{\sqrt{2}}(|00\rangle + |11\rangle)$, $|\Phi^-\rangle = \frac{1}{\sqrt{2}}(|00\rangle - |11\rangle)$, $|\Psi^+\rangle = \frac{1}{\sqrt{2}}(|01\rangle + |10\rangle)$, and $|\Psi^-\rangle = \frac{1}{\sqrt{2}}(|01\rangle - |10\rangle)$. They can be similarly generated, using an appropriate quantum circuit (the reader is encouraged to try to figure out the relevant circuits to generate the above states).

To simplify diagrams which follow in the book, we will use a squiggly or wavy line to represent the circuit in Figure 2.10, as shown in Figure 2.11.

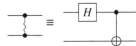

Figure 2.11 Simplified diagram for entangling a pair of qubits.

Note that to generate a cat-like state $\alpha |00\rangle + \beta |11\rangle$, we can use a CNOT gate with control $\alpha |0\rangle + \beta |1\rangle$ and target $|0\rangle$ as we have seen in Figure 2.9.

For three qubits, one can consider the Greenberger–Horne–Zeilinger (GHZ) state:

$$|GHZ\rangle = \frac{1}{\sqrt{2}} |000\rangle + \frac{1}{\sqrt{2}} |111\rangle$$

Such a state can be generated via a circuit of the form shown in Figure 2.12 applied to the three qubit state $|000\rangle$. Note that we can use the circuit shown in Figure 2.13 to generate the state $\alpha |000\rangle + \beta |111\rangle$ in Figure 2.13. And in general, for n qubits, we have the generalized GHZ states of the form: $|GHZ_n\rangle = \frac{1}{\sqrt{2}}|0\rangle^{\otimes n} + \frac{1}{\sqrt{2}}|1\rangle^{\otimes n}$.[22]

Figure 2.12 Simplified diagram for entangling three qubits.

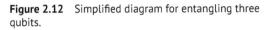
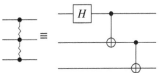

---

21 Named after Albert Einstein, Boris Podolsky, and Nathan Rosen.

22 Note that we use the notation $A^{\otimes n}$ as short for $\overbrace{A \otimes A \otimes \cdots \otimes A}^{n}$, i.e. tensor product with itself multiple times.

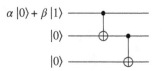
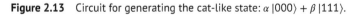

Figure 2.13 Circuit for generating the cat-like state: $\alpha \, |000\rangle + \beta \, |111\rangle$.

Such a state can be generated via a circuit shown in Figure 2.14a applied to the n qubit state $|0\rangle^{\otimes n}$ using $O(n)$ steps, or by a more efficient implementation using $O(\log n)$ steps, illustrated with eight qubits in Figure 2.14b, where we observe that multiple qubits can be acted on in the same step, in order to reduce the total number of steps, thereby increasing efficiency. The number of such steps in a circuit is

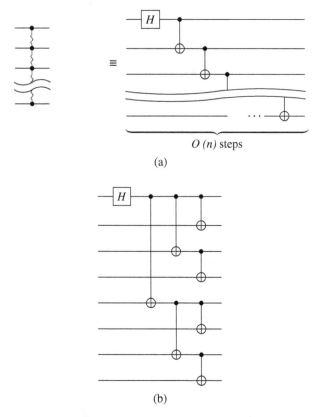

$O(n)$ steps

(a)

(b)

Figure 2.14 Circuits to generate the generalized GHZ state. (a) Circuit to generate the generalized GHZ state in a linear number of steps. (b) Circuit to generate the generalized GHZ state in a logarithmic number of steps.

also called the *circuit depth*.[23] A compiler could be written to optimize a quantum circuit, i.e. generate a quantum circuit of lower depth that performs the equivalent operations. A circuit similar to and generalizing the one in Figure 2.13 can be used to generate the n-qubit cat-like state: $|CAT_n\rangle = \alpha|0\rangle^{\otimes n} + \beta|1\rangle^{\otimes n}$.

The GHZ state involves more than two qubits and so is a *multipartite entanglement*, e.g. where the entangled state is shared by multiple qubits, one qubit on each node. Interestingly, it has been observed experimentally as early as 1998 [Bouwmeester et al., 1999]. It is also noted in Illiano et al. [2022] that an EPR pair between the two so-called *winning nodes* can be "extracted" from a (generalized) GHZ state shared among multiple nodes via measurements at the *losing nodes* (i.e. the rest of the nodes which are not the winning nodes). For example, suppose we have a GHZ state, with three qubits, one on each node. Then, suppose we first apply H and then do a measurement on the third qubit (the losing node), i.e. we apply the following operation:

$$(I \otimes I \otimes H)|GHZ\rangle$$

$$= (I \otimes I \otimes H)(\frac{1}{\sqrt{2}}|000\rangle + \frac{1}{\sqrt{2}}|111\rangle)$$

$$= \frac{1}{\sqrt{2}}|00\rangle H|0\rangle + \frac{1}{\sqrt{2}}|11\rangle H|1\rangle$$

$$= \frac{1}{\sqrt{2}}|00\rangle (\frac{1}{\sqrt{2}}|0\rangle + \frac{1}{\sqrt{2}}|1\rangle) + \frac{1}{\sqrt{2}}|11\rangle (\frac{1}{\sqrt{2}}|0\rangle - \frac{1}{\sqrt{2}}|1\rangle)$$

$$= \frac{1}{\sqrt{2}}(|00\rangle + |11\rangle) \otimes \frac{1}{\sqrt{2}}|0\rangle + \frac{1}{\sqrt{2}}(|00\rangle - |11\rangle) \otimes \frac{1}{\sqrt{2}}|1\rangle$$

$$= |\Phi^+\rangle \otimes \frac{1}{\sqrt{2}}|0\rangle + |\Phi^-\rangle \otimes \frac{1}{\sqrt{2}}|1\rangle$$

and then when we measure the third qubit, we get either (i) the resulting state $|0\rangle$ in which case the first two nodes are left with the shared entangled EPR pair $|\Phi^+\rangle$ or (ii) the resulting state $|1\rangle$ in which case the first two nodes are left with the shared entangled EPR pair $|\Phi^-\rangle$, i.e. either way, two nodes are left with an EPR pair. Note that measuring the third qubit is a local operation on the third node containing the qubit, and classical communication might be used to inform the first two nodes that measurement has been done and so that the winning nodes now share an EPR pair (perhaps informing them which Bell state they now share).

---

23 More precisely, the circuit depth is the longest path between the input state and the output state, with each gate counting as a unit. Some examples are at: https://quantumcomputing .stackexchange.com/questions/5769/how-to-calculate-circuit-depth-properly [last accessed: 28/10/2022]

More generally, with a generalized GHZ state shared among n nodes, one qubit on each node, and suppose we want to generate an EPR pair between nodes i and j we can first apply the following operation: $U = U_1 \otimes \cdots \otimes U_n$, where $U_i = U_j = I$, and $U_k = H$ for all $k \neq i$ and $k \neq j$:

$$(U_1 \otimes \cdots \otimes U_n)|GHZ_n\rangle$$

$$= (U_1 \otimes \cdots \otimes U_n)(\frac{1}{\sqrt{2}}|0\rangle^{\otimes n} + \frac{1}{\sqrt{2}}|1\rangle^{\otimes n})$$

$$= \frac{1}{\sqrt{2}}|0\rangle_i|0\rangle_j \otimes H^{\otimes(n-2)}|0\rangle^{\otimes(n-2)} + \frac{1}{\sqrt{2}}|1\rangle_i|1\rangle_j \otimes H^{\otimes(n-2)}|1\rangle^{\otimes(n-2)}$$

(reordering the qubits so that qubits i and j are written to be on the left)

$$= \frac{1}{\sqrt{2}}|0\rangle_i|0\rangle_j \otimes \frac{1}{\sqrt{2^{n-2}}} \sum_{x \in \{0,1\}^{n-2}} |x\rangle$$

$$+ \frac{1}{\sqrt{2}}|1\rangle_i|1\rangle_j \otimes \frac{1}{\sqrt{2^{n-2}}} \sum_{y \in \{0,1\}^{n-2}} (-1)^{parity(y)} |y\rangle$$

(where $parity(y) = parity(y_1 \ldots y_{n-2}) = y_1 \oplus \cdots \oplus y_{n-2}$,
and $\oplus$ is addition modulo 2)

$$= |\Phi^+\rangle_{ij} \otimes \frac{1}{\sqrt{2^{n-2}}} \sum_{x \in evenp} |x\rangle + |\Phi^-\rangle_{ij} \otimes \frac{1}{\sqrt{2^{n-2}}} \sum_{y \in oddp} |y\rangle$$

(where $evenp = \{x'|x' \in \{0,1\}^{n-2}$ and parity $(x') = 0\}$, and
$oddp = \{y'|y' \in \{0,1\}^{n-2}$ and parity $(y') = 1\}$)

and then when we measure all the qubits except for i and j, we get either (i) a resulting state $|x\rangle$ where $x \in evenp$ in which case the two nodes i and j are left with the shared entangled EPR pair $|\Phi^+\rangle$ or (ii) a resulting state $|y\rangle$ where $y \in oddp$ in which case the two nodes i and j are left with the shared entangled EPR pair $|\Phi^-\rangle$, i.e. either way, the two qubits i and j are left with an EPR pair. This has potential applications for the future quantum Internet where suppose that a multipartite entanglement can be shared among n nodes, then by the nodes applying the appropriate operations and measurements, an EPR pair viewed as a resource can be extracted for any two winning nodes (e.g. the two nodes chosen to share an EPR pair for that round). Also, it is possible to generate GHZ states distributed over multiple nodes from fusing bipartite entangled pairs (distributed over pairs of nodes), e.g. create an n-qubit GHZ state from fusing $(n-1)$ Bell pairs (as suggested by Figure 2.14) – we will see this in Chapter 4, when we see how Bell pairs (distributed across nodes) are used for non-local CNOT operations – a discussion of more efficient protocols for creating GHZ states from Bell pairs is in de Bone et al. [2020].

Apart from the GHZ state, there is a three-qubit (or tripartite) entangled state called the W state of the form:

$$|W\rangle = \frac{1}{\sqrt{3}}(|001\rangle + |010\rangle + |100\rangle)$$

It has been noted by Dür et al. [2000] that any tripartite entangled state can be converted using Local Operations and Classical Communication (LOCC) (e.g. local gates/operations on each node holding one qubit and classical communications between nodes) into one of two standard forms, the GHZ state or the W state (not both), and that if a state $|\psi\rangle$ is convertible (via LOCC) into a GHZ state and another state $|\phi\rangle$ into the W state, then $|\psi\rangle$ cannot be converted into $|\phi\rangle$ and conversely. This means that GHZ and W are in this sense two very different kinds of entangled states.

An interesting observation about the W state is that it is more robust to errors than $|GHZ\rangle$, in the following sense. Suppose we have the W state shared among three qubits, and one of the qubits is measured, say the last qubit, then with probability 1/3 we get $|1\rangle$ and the other qubits end up in state $|00\rangle$ (corresponding to $|001\rangle$). But with probability 2/3 we get $|0\rangle$, and the other qubits end up in an entangled state: $\frac{1}{\sqrt{2}}(|01\rangle + |10\rangle)$ (corresponding to the superposition of $|010\rangle$ and $|100\rangle$). But with the GHZ state, on measuring the last qubit, we get either $|0\rangle$ or $|1\rangle$, and no chance of the remaining two qubits being in an entangled state. Hence, three qubits, if sharing a W state and one qubit is measured (or "lost" as a result of some interaction with the environment), there is still a probability of 2/3 that the remaining two qubits share an entangled state (retaining the resource of entanglement), but if sharing a GHZ state and one qubit "lost," no resource (entanglement) can be retained. This fact might have significance for the quantum network, in terms of robustness.

The generalized W_n state is of the form:

$$|W_n\rangle = \frac{1}{\sqrt{n}}(|100...000\rangle + |010...000\rangle + \cdots + |000...010\rangle + |000...001\rangle)$$

with each label being n-bit strings. Similar arguments about robustness can be considered for the generalized case, compared to $|GHZ_n\rangle$.

2.4 Teleportation and Superdense Coding

Quantum teleportation was mentioned earlier, and we now provide the mathematical model of this technique of transferring a quantum state from one node to another.

A key resource required for quantum teleportation is entanglement. Consider the problem of transferring a quantum state $|\psi\rangle = \alpha|0\rangle + \beta|1\rangle$ from one node to

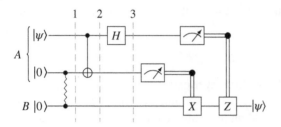

Figure 2.15 A circuit for teleporting a quantum state from A to B.

another. Figure 2.15 shows the circuit for teleporting a quantum state ψ from a node A to B using an EPR pair. Node A has two qubits, one whose state $|\psi\rangle$ is to be transferred to B, and another qubit sharing an entangled state with a qubit at node B. The wavy arrow is the circuit as shown in Figure 2.11. Toward the end of the computations, node A sends two classical bits to B which will determine whether the X and Z gates will be applied, and the result is that B's qubit takes on the state of A's qubit $|\psi\rangle$. Note that A's qubit no longer has this state since it is measured, which is consistent with the no-cloning theorem that, generally, there is no quantum circuit that can copy any unknown qubit state.

Let us follow the calculations to see how it works. At slice 1, we have the following state of the three qubits (two at A, denoted by subscripts for qubits 1 and 2, and one at B denoted by a subscript for qubit 3), i.e., we use subscripts to label the qubits where useful to do so:

$$|\psi\rangle_1 \otimes \frac{1}{\sqrt{2}}(|00\rangle_{23} + |11\rangle_{23})$$

Then after applying the CNOT gate, at slice 2, we have:

$$(CNOT_{12} \otimes I_3)\left(|\psi\rangle_1 \otimes \frac{1}{\sqrt{2}}(|00\rangle_{23} + |11\rangle_{23})\right)$$

$$= (CNOT_{12} \otimes I_3)\left((\alpha|0\rangle_1 + \beta|1\rangle_1) \otimes \frac{1}{\sqrt{2}}(|00\rangle_{23} + |11\rangle_{23})\right)$$

$$= \frac{1}{\sqrt{2}}(CNOT_{12} \otimes I_3)(\alpha|000\rangle + \beta|100\rangle + \alpha|011\rangle + \beta|111\rangle)$$

$$= \frac{1}{\sqrt{2}}(\alpha|000\rangle + \beta|110\rangle + \alpha|011\rangle + \beta|101\rangle)$$

$$= \frac{1}{\sqrt{2}}(\alpha|0\rangle_1(|00\rangle_{23} + |11\rangle_{23}) + \beta|1\rangle_1(|10\rangle_{23} + |01\rangle_{23}))$$

Then, we apply the H gate to the first qubit to get the following state at slice 3:

$$= (H_1 \otimes I_2 \otimes I_3) \frac{1}{\sqrt{2}} \left(\alpha |0\rangle_1 (|00\rangle_{23} + |11\rangle_{23}) + \beta |1\rangle_1 (|10\rangle_{23} + |01\rangle_{23}) \right)$$

$$= \frac{1}{\sqrt{2}} \left(\alpha H |0\rangle_1 (|00\rangle_{23} + |11\rangle_{23}) + \beta H |1\rangle_1 (|10\rangle_{23} + |01\rangle_{23}) \right)$$

$$= \frac{1}{\sqrt{2}} \left(\alpha \frac{1}{\sqrt{2}} (|0\rangle_1 + |1\rangle_1)(|00\rangle_{23} + |11\rangle_{23}) \right.$$

$$\left. + \beta \frac{1}{\sqrt{2}} (|0\rangle_1 - |1\rangle_1)(|10\rangle_{23} + |01\rangle_{23}) \right)$$

$$= \frac{1}{2} (\alpha |000\rangle + \alpha |100\rangle + \alpha |011\rangle + \alpha |111\rangle + \beta |010\rangle$$

$$- \beta |110\rangle + \beta |001\rangle - \beta |101\rangle)$$

$$= \frac{1}{2} (\alpha |000\rangle + \beta |001\rangle + \alpha |011\rangle + \beta |010\rangle + \alpha |100\rangle$$

$$- \beta |101\rangle + \alpha |111\rangle - \beta |110\rangle)$$

$$= \frac{1}{2} (|00\rangle \otimes (\alpha |0\rangle + \beta |1\rangle) + |01\rangle \otimes (\alpha |1\rangle + \beta |0\rangle)$$

$$+ |10\rangle \otimes (\alpha |0\rangle - \beta |1\rangle) + |11\rangle \otimes (\alpha |1\rangle - \beta |0\rangle))$$

$$= \frac{1}{2} \left(|00\rangle \otimes X^0 Z^0 (\alpha |0\rangle + \beta |1\rangle) + |01\rangle \otimes XZ^0 (\alpha |0\rangle + \beta |1\rangle) \right.$$

$$\left. + |10\rangle \otimes X^0 Z (\alpha |0\rangle + \beta |1\rangle) + |11\rangle \otimes XZ (\alpha |0\rangle + \beta |1\rangle) \right)$$

(noting that $X^0 = Z^0 = I$)

$$= \frac{1}{2} \sum_{p,q \in \{0,1\}} |pq\rangle_{12} \otimes [X^q Z^p (\alpha |0\rangle + \beta |1\rangle)]_3$$

Note that, suppose now, the measurement results are p and q corresponding to the state of the first two qubits being $|pq\rangle$, as in the circuit, A sends the (classical) bits p and q to B, and B can apply $Z^p X^q$ to the third qubit to get:

$$(Z^p X^q)[X^q Z^p (\alpha |0\rangle + \beta |1\rangle)] = Z^p X^q X^q Z^p (\alpha |0\rangle + \beta |1\rangle) = \alpha |0\rangle + \beta |1\rangle$$

since $XX = ZZ = I$, and so, the qubit at B has taken on the required state. Hence, the (communication) resources required for teleporting the quantum state are the communication of two classical bits and one EPR pair (which is "consumed" in the teleportation). One can effectively transfer a quantum state encoded in a transmitted photon, but the above method shows that the transfer of a quantum state can be done via transmitting classical bits as long as there is one EPR pair available and the ability to perform certain quantum operations.

Instead of transmitting two bits to communicate one qubit of information, one can transmit one qubit to communicate two bits of information, as long as there

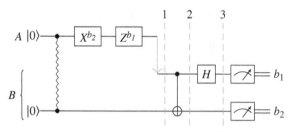

Figure 2.16 A circuit for transmitting two classical bits of information by sending one qubit (indicated by an arrow) from A to B.

is an EPR pair available to be "consumed," using a protocol known as *superdense coding*, illustrated in Figure 2.16. We start with an EPR pair shared by A and B. Then, for a choice of b_1 and b_2 depending on which two bits A wants to send, after applying the gates X^{b_2} and Z^{b_1} to the qubit at A, we get, at slice 1: $(Z^{b_1} X^{b_2} \otimes I)$ $\frac{1}{\sqrt{2}}(|00\rangle + |11\rangle)$. Subsequent operations in the circuit for different choices of b_1 and b_2 are given in the following table:

b_1	b_2	At slice 1	At slice 2	At slice 3
0	0	$\frac{1}{\sqrt{2}}(|00\rangle + |11\rangle)$	$\frac{1}{\sqrt{2}}(|00\rangle + |10\rangle)$	$\frac{1}{2}(|00\rangle + |10\rangle + |00\rangle - |10\rangle) = |00\rangle$
0	1	$\frac{1}{\sqrt{2}}(|10\rangle + |01\rangle)$	$\frac{1}{\sqrt{2}}(|11\rangle + |01\rangle)$	$\frac{1}{2}(|01\rangle - |11\rangle + |01\rangle + |11\rangle) = |01\rangle$
1	0	$\frac{1}{\sqrt{2}}(|00\rangle - |11\rangle)$	$\frac{1}{\sqrt{2}}(|00\rangle - |10\rangle)$	$\frac{1}{2}(|00\rangle + |10\rangle - |00\rangle + |10\rangle) = |10\rangle$
1	1	$\frac{1}{\sqrt{2}}(-|10\rangle + |01\rangle)$	$\frac{1}{\sqrt{2}}(-|11\rangle + |01\rangle)$	$\frac{1}{2}(-|01\rangle + |11\rangle + |01\rangle + |11\rangle) = |11\rangle$

Upon measurement, note that the bits obtained are indeed b_1 and b_2.

2.5 Summary

In this chapter, we have looked at basic ideas of quantum information and quantum computation, including concepts of quantum entanglement and teleportation. We have introduced only the main concepts required for the rest of the book, and as we proceed, we will introduce further concepts as needed. In Section 2.6, we point out further learning resources for one new to quantum computing.

2.6 Further Reading and Resources on Quantum Computing

For someone learning quantum information and quantum computation, the book that cannot be ignored is the "quantum computing bible," i.e. Isaac Chuang and

Michael Nielsen's *Quantum Computation and Quantum Information*. A number of books on quantum computing also helps, including A. Yu. Kitaev, A.H. Shen, and M.N. Vyalyi's book on *Classical and Quantum Computation*, Noson S. Yanofsky and Mirco A. Mannucci's book *Quantum Computing for Computer Scientists*, Eleanor Rieffel and Wolfgang Polak's *Quantum Computing: A Gentle Introduction*, Dan C. Marinescu and Gabriela M. Marinescu's *Approaching Quantum Computing*, Phillip Kaye, Raymond Laflamme, and Michele Mosca's *An Introduction to Quantum Computing*, N. David Mermin's *Quantum Computer Science: An Introduction* and several others. If the above books are too challenging, David McMahon's book *Quantum Computing Explained* is a good start where a lot of the calculations are worked out. Also, recent years have seen more introductory books available. Two freely available books for beginners (especially with some IT background but not physics) are Thomas G. Wong's *Introduction to Classical and Quantum Computing*[24] and *Quantum Computing for the Quantum Curious* by Ciaran Hughes, Joshua Isaacson, Anastasia Perry, Ranbel F. Sun, and Jessica Turner.[25] Then, particularly for programmers, Robert Hundt's *Quantum Computing for Programmers* is excellent. There are a number of excellent online videos (and video series) on quantum computing. For example, one I have used is Umesh Vazirani's lectures on introductory quantum computing available online.[26] For someone interested in quantum mechanics, Barton Zwiebach's lecture series is enlightening.[27] IBM's online textbook on the quantum computing platform Qiskit provides a good introduction.[28] A short online course on Quantum Computing offered by the Massachusetts Institute of Technology, since several years ago, might still be offered perhaps in newer versions.[29] By the time this book gets into your hands, there would have been many more resources available online.

References

Sebastian de Bone, Runsheng Ouyang, Kenneth Goodenough, and David Elkouss. Protocols for creating and distilling multipartite GHZ states with Bell pairs. *IEEE Transactions on Quantum Engineering*, 1:1–10, 2020. doi: 10.1109/TQE.2020.3044179.

---

24 https://www.thomaswong.net/introduction-to-classical-and-quantum-computing-1e3p.pdf [last accessed: 9/10/2022]
25 https://link.springer.com/book/10.1007/978-3-030-61601-4) [last accessed: 9/10/2022]
26 See https://www.youtube.com/channel/UCq9B8tT3oXl8BSyaoBPQXQw/videos [last accessed: 7/11/2022]
27 https://ocw.mit.edu/courses/8-04-quantum-physics-i-spring-2016 [last accessed: 7/11/2022]
28 See the Qiskit textbook https://qiskit.org/learn [last accessed: 7/11/2022].
29 See https://learn-xpro.mit.edu/quantum-computing [last accessed: 7/11/2022]

Dik Bouwmeester, Jian-Wei Pan, Matthew Daniell, Harald Weinfurter, and Anton Zeilinger. Observation of three-photon greenberger-horne-zeilinger entanglement. *Physical Review Letters*, 82:1345–1349, Feb 1999. doi: 10.1103/PhysRevLett.82.1345.

W. Dür, G. Vidal, and J. I. Cirac. Three qubits can be entangled in two inequivalent ways. *Physical Review A*, 62:062314, Nov 2000. doi: 10.1103/PhysRevA.62.062314.

Ciaran Hughes, Joshua Isaacson, Anastasia Perry, Ranbel F. Sun, and Jessica Turner. *Creating Superposition: The Beam Splitter*, pages 17–28. Springer International Publishing, Cham, 2021a.

Ciaran Hughes, Joshua Isaacson, Anastasia Perry, Ranbel F. Sun, and Jessica Turner. *What Is a Qubit?* pages 7–16. Springer International Publishing, Cham, 2021b.

Jessica Illiano, Michele Viscardi, Seid Koudia, Marcello Caleffi, and Angela Sara Cacciapuoti. Quantum Internet: From Medium Access Control to Entanglement Access Control. May 2022.

Louis H. Kauffman and Sam J. Lomonaco Jr., Quantum diagrams and quantum networks. In Eric Donkor, Andrew R. Pirich, Howard E. Brandt, Michael R. Frey, Samuel J. Lomonaco Jr., and John M. Myers, editors, *Quantum Information and Computation XII*, volume 9123, page 91230P. International Society for Optics and Photonics, SPIE, 2014.

D. Leibfried, E. Knill, S. Seidelin, J. Britton, R. B. Blakestad, J. Chiaverini, D. B. Hume, W. M. Itano, J. D. Jost, C. Langer, R. Ozeri, R. Reichle, and D. J. Wineland. Creation of a six-atom 'Schrödinger cat'state. *Nature*, 438(7068):639–642, 2005.

Michael A. Nielsen and Isaac L. Chuang. *Quantum Computation and Quantum Information*. Cambridge University Press, 2000.

3

Distributed Quantum Computing – Classical and Quantum

Early discussions on distributed quantum computing focused on communication complexity, that is, the number of bits (or qubits) in messages exchanged in a distributed computation involving at least two parties. The aim was to see if entanglement provided a way to reduce the communication complexity of (classical or quantum) distributed algorithms – reducing communication among nodes is a key goal of many distributed computing algorithms since such a reduction can improve efficiency, reduce the coupling among nodes, and reduce the use of communication resources.

We first consider three of these algorithms: the distributed three-party product problem, the distributed Deutsch–Jozsa promise problem, and the distributed intersection problem. While one might argue that these are not quite practical distributed quantum computing algorithms, we will see how quantum information can enable fewer bits of communication to be used for a distributed computing solution, as opposed to using just classical information.

Then, we will also look at an amazing result that some distributed computation problems (with certain constraints and requirements) cannot be solved classically, but can be solved with the use of quantum information, in particular entanglement. In fact, the ability to solve such a problem within the constraints and requirements might be considered a test of whether there is indeed quantum entanglement!

Thereafter, we will look at a number of quantum protocols, which seem rather theoretical, but many have practical implications, and involve multiple parties (or nodes), and hence, distributed in nature. Quantum protocols we will look at include quantum coin flipping, quantum leader election, quantum key distribution, quantum anonymous broadcasting, quantum voting, quantum Byzantine Generals, quantum secret sharing, and quantum oblivious transfer (OT).

From Distributed Quantum Computing to Quantum Internet Computing: An Introduction,
First Edition. Seng W. Loke.
© 2024 The Institute of Electrical and Electronics Engineers, Inc. Published 2024 by John Wiley & Sons, Inc.

3.1 The Power of Entanglement for Distributed Computing

We introduced the notion of entanglement in Chapter 2. As noted by Cleve and Buhrman [1997], no communication actually happens when measuring entangled particles, or to put this another way, two parties sharing entangled particles cannot use this entanglement to communicate, in this sense: suppose A and B each has an ebit, i.e. sharing an entangled pair of qubits, say $\frac{1}{\sqrt{2}}(|00\rangle + |11\rangle)$ then when A measures its qubit, and obtains x (either 0 or 1), then B also obtains the same value x, but since A cannot determine before measurement what x will be, A cannot choose what to communicate to B.

But a question is whether the entanglement between A and B can, in some way, reduce the communication between A and B when performing some computation that involves data on A and on B. We will see that the answer is yes.

3.1.1 Enabling Distributed Computations with Fewer Bits of Communication

To make precise what we mean by reducing communication when performing certain distributed computations, we first introduce the notion of *communication complexity*. Following the exposition in Cleve and Buhrman [1997], suppose we have two different machines A and B, with A having an n-bit string $a = a_1 \dots a_n$ and B having an n-bit string $b = b_1 \dots b_n$, and suppose the intended distributed computation is the sum (modulo 2) function $f : \{0,1\}^n \times \{0,1\}^n \to \{0,1\}$, which requires the inputs of A and B:

$$f(a,b) = a_1 \oplus \cdots \oplus a_n \oplus b_1 \oplus \cdots \oplus b_n$$

where $\oplus$ is the addition modulo 2 operation. The minimum number of bits that B needs to send to A for A to compute $f(a,b)$ is just one bit, namely, the computed result $b_1 \oplus \cdots \oplus b_n$. As an example, consider another computation, i.e. inner product (modulo 2) defined as follows:

$$f(a,b) = (a_1 \cdot b_1) \oplus \cdots \oplus (a_n \cdot b_n)$$

Then, for A to compute $f(a,b)$, B has no other way but to send all its n bits, i.e. the minimum number of bits B has to send to A in this case is n. The minimum number of classical bits that must be sent for A to compute the function f is the *communication complexity* of f. For example, for the sum (modulo 2) function, the communication complexity is 1 whereas for inner product it is n.

We will look at three problems where we see the quantum advantage with respect to communication complexity.

3.1.1.1 The Distributed Three-Party Product Problem

Let's consider an example from Cleve and Buhrman [1997], where in the classical case (without entanglement), three classical bits of communication is required at minimum, whereas with entanglement among machines, only two classical bits of communication is enough, that is, entanglement can reduce the communication complexity. We have A, B, and C with n-bit strings x^A, x^B, and x^C, respectively,

with the condition: $x^A \oplus x^B \oplus x^C = \overbrace{1 \ldots 1}^{n} = \mathbf{1}$, where the $\oplus$ here is bitwise addition (modulo 2), and the function f defined as $f(a,b,c) = (a_1 \cdot b_1 \cdot c_1) \oplus \cdots \oplus (a_n \cdot b_n \cdot c_n)$, that is, we want A to compute the value

$$f(x^A, x^B, x^C) = (x_1^A \cdot x_1^B \cdot x_1^C) \oplus \cdots \oplus (x_n^A \cdot x_n^B \cdot x_n^C)$$

Note that the condition above on n-bit strings x^A, x^B, and x^C makes this problem not quite a "generalization" of the inner product we saw earlier (which doesn't have such a condition on the inputs).

We assume that A, B, and C share an entangled collection of $3n$ qubits, with each having n qubits, i.e. there are n triples of entangled qubits, each triple denoted by $\psi_i^A \psi_i^B \psi_i^C$, with $i \in \{1, \ldots, n\}$, in an entangled state of the form: $\rho_{ABC} = \frac{1}{2}(|001\rangle + |010\rangle + |100\rangle - |111\rangle)$. Note that this state can be obtained from the Greenberger–Horne–Zeilinger (GHZ) state $\frac{1}{\sqrt{2}}(|000\rangle + |111\rangle)$ via local unitary (or LU) operations (i.e. operations done on each machine locally), i.e. up to LU operations, it is equivalent to the GHZ state – see **Aside**.

Aside: There could be different possible operations; one is as follows:

1. Starting with $\frac{1}{\sqrt{2}}(|000\rangle + |111\rangle)$, we apply the operation $Z \otimes Z \otimes Z$:

$$(Z \otimes Z \otimes Z)\frac{1}{\sqrt{2}}(|000\rangle + |111\rangle) = \frac{1}{\sqrt{2}}(|000\rangle - |111\rangle)$$

Note that an operation such as $I \otimes I \otimes Z$ can be used as well.

2. Apply the operation $H \otimes H \otimes H$:

$$(H \otimes H \otimes H)\frac{1}{\sqrt{2}}(|000\rangle - |111\rangle)$$

$$= \frac{1}{\sqrt{2}}(H|0\rangle H|0\rangle H|0\rangle - H|1\rangle H|1\rangle H|1\rangle)$$

$$= \frac{1}{4}((|000\rangle + |001\rangle + |010\rangle + |011\rangle + |100\rangle + |101\rangle + |110\rangle + |111\rangle)$$

$$- (|000\rangle - |001\rangle - |010\rangle + |011\rangle - |100\rangle + |101\rangle + |110\rangle - |111\rangle))$$

$$= \frac{1}{4}(|000\rangle + |001\rangle + |010\rangle + |011\rangle + |100\rangle + |101\rangle + |110\rangle + |111\rangle$$
$$- |000\rangle + |001\rangle + |010\rangle - |011\rangle + |100\rangle - |101\rangle - |110\rangle + |111\rangle)$$
$$= \frac{1}{2}(|001\rangle + |010\rangle + |100\rangle + |111\rangle)$$

3. Apply the operation $\sqrt{Z} \otimes \sqrt{Z} \otimes \sqrt{Z}$, where $\sqrt{Z} = \begin{bmatrix} 1 & 0 \\ 0 & i \end{bmatrix}$, i.e. $\sqrt{Z}|0\rangle = |0\rangle$ and $\sqrt{Z}|1\rangle = i|1\rangle$ (and $\sqrt{Z}\sqrt{Z} = Z$) – note that we also call the *square root of Z* gate the *P* gate later in Chapter 5):

$$\left(\sqrt{Z} \otimes \sqrt{Z} \otimes \sqrt{Z} \right) \frac{1}{2}(|001\rangle + |010\rangle + |100\rangle + |111\rangle)$$
$$= \frac{1}{2}(i|001\rangle + i|010\rangle + i|100\rangle + i(i)^2|111\rangle)$$
$$= i\frac{1}{2}(|001\rangle + |010\rangle + |100\rangle - |111\rangle)$$

and ignoring the global phase i, we have $\frac{1}{2}(|001\rangle + |010\rangle + |100\rangle - |111\rangle)$ as required.

Now we run the following algorithm with A, B, and C, where H is the Hadamard transform mapping $|0\rangle$ to $\frac{1}{\sqrt{2}}(|0\rangle + |1\rangle)$ and $|1\rangle$ to $\frac{1}{\sqrt{2}}(|0\rangle - |1\rangle)$), and measurements are in the standard (or computational) basis:

---

0. for each $p \in \{A, B, C\}$:
1.   **for each** $i \in \{1, \dots, n\}$:
2.   **if** $x_i^p = 0$ **then**
3.   **apply** H **to** ψ_i^p
4.   **else**
5.   **do nothing to** ψ_i^p
6.   **measure** ψ_i^p **yielding bit** s_i^p
7. $s^p := s_1^p \oplus \dots \oplus s_n^p$

Then, B and C send their results (i.e. s^B and s^C) to A, which already has s^A, and computes the value $s^A \oplus s^B \oplus s^C$, and since we have the condition $x^A \oplus x^B \oplus x^C = 1$,

$$s^A \oplus s^B \oplus s^C = f(x^A, x^B, x^C)$$

and A has succeeded with only two classical bits being sent to it. Let us see why the above equation holds.

First, we observe that for each i:

$$s_i^A \oplus s_i^B \oplus s_i^C = x_i^A \cdot x_i^B \cdot x_i^C$$

This is because since $x^A \oplus x^B \oplus x^C = 1$, i.e. $x_i^A \oplus x_i^B \oplus x_i^C = 1$, there are only four possible combinations of values for x_i^A, x_i^B, x_i^C, which are 0, 0, 1 or 0, 1, 0 or 1, 0, 0 or 1, 1, 1 (only odd number of ones), and we can see from the following table, that, for each i, for each possible value of x_i^A, x_i^B, x_i^C, we have that $s_i^A \oplus s_i^B \oplus s_i^C = x_i^A \cdot x_i^B \cdot x_i^C$.

$x_i^A \cdot x_i^B \cdot x_i^C$	State after line 5 in the algorithm	$s_i^A \oplus s_i^B \oplus s_i^C$
$0 \cdot 0 \cdot 1 = 0$	$H \otimes H \otimes I(\rho_{ABC}) = \frac{1}{2}(|011\rangle + |101\rangle + |000\rangle - |110\rangle)$	0
$0 \cdot 1 \cdot 0 = 0$	$H \otimes I \otimes H(\rho_{ABC}) = \frac{1}{2}(|011\rangle + |110\rangle + |000\rangle - |101\rangle)$	0
$1 \cdot 0 \cdot 0 = 0$	$I \otimes H \otimes H(\rho_{ABC}) = \frac{1}{2}(|101\rangle + |110\rangle + |000\rangle - |011\rangle)$	0
$1 \cdot 1 \cdot 1 = 1$	$I \otimes I \otimes I(\rho_{ABC}) = \frac{1}{2}(|001\rangle + |010\rangle + |100\rangle - |111\rangle)$	1

Note that, for each i, the state of the triple of qubits after selectively applying the Hadamard gates (middle column in the table) depends on which of the four combinations of values for x_i^A, x_i^B, x_i^C are present.

Following from above, we have:

$$\begin{aligned}
s^A \oplus s^B \oplus s^C &= (s_1^A \oplus \cdots \oplus s_n^A) \oplus (s_1^B \oplus \cdots \oplus s_n^B) \oplus (s_1^C \oplus \cdots \oplus s_n^C) \\
&= (s_1^A \oplus s_1^B \oplus s_1^C) \oplus \cdots \oplus (s_n^A \oplus s_n^B \oplus s_n^C) \\
&= (x_1^A \cdot x_1^B \cdot x_1^C) \oplus \cdots \oplus (x_n^A \cdot x_n^B \cdot x_n^C) \\
&= f(x^A, x^B, x^C)
\end{aligned}$$

which is the value computed by A. As explained in Cleve and Buhrman [1997], without shared entangled bits, two classical bits of communication among A, B, and C are insufficient for A to compute $f(a, b, c)$, and three classical bits are required to be communicated. In an interesting way, the shared qubits seem to be "taking the place" of one classical bit of communication: n entangled triples of qubits plus two classical bits of communication is equivalent to three classical bits of communication, for this problem!

3.1.1.2 The Distributed Deutsch–Jozsa Promise Problem

Let us consider another problem based on the one in Buhrman et al. [2010], related to another problem (and its solution/algorithm) called the Deutsch–Jozsa algorithm,[1] as follows. Suppose we have A and B, and A has an n-bit string $a = a_1 \ldots a_n$

---

1 Note that the original Deutsch–Jozsa problem is to determine if a given function (say $f : \{0,1\}^n \to \{0,1\}$) is constant (returns 0 on all its inputs or 1 on all its inputs) or balanced (returns 0 on half its inputs and 1 on the other half of its inputs) – and we are promised that the function is either one or the other. A solution is at https://en.wikipedia.org/wiki/Deutsch-Jozsa_algorithm.

and B has an n-bit string $b = b_1 \dots b_n$. The problem is to determine whether (1) $a = b$, or (2) a and b are different in exactly $n/2$ positions, for n an even number (here, we also assume that n is a power of 2), and this is a *promise problem* which means that it does not apply to all pairs of strings $(x, y) \in \{0, 1\}^n \times \{0, 1\}^n$, but only to those where either (1) or (2) is true (clearly, both cannot be true for a given pair of strings).

An algorithm to solve this problem is as follows, assuming $n = 2^l$ (or $l = \log_2 n$):

1. B, using b and l $|0\rangle$-qubits, prepares the state $\frac{1}{\sqrt{n}} \sum_{i=0}^{n-1} (-1)^{b_i} |i\rangle$; note that the idea here is to encode the n bits of information in b in the "signs" (i.e. $(-1)^{b_i}$) of the n components of an l-qubit state (since $l \ll 2^l$, we are attempting to "get away" with transmitting fewer bits). This can be done as follows:

    (i) apply $H^{\otimes l}$ on l $|0\rangle$-qubits to obtain the state $\frac{1}{\sqrt{n}} \sum_{i=0}^{n-1} |i\rangle$; for example, for $n = 2^3$, $H \otimes H \otimes H |000\rangle = \frac{1}{\sqrt{2}}(|0\rangle + |1\rangle) \otimes \frac{1}{\sqrt{2}}(|0\rangle + |1\rangle) \otimes \frac{1}{\sqrt{2}}(|0\rangle + |1\rangle)$

    $= \frac{1}{\sqrt{2^3}} \sum_{i=0}^{2^3-1} |i\rangle$

    (ii) assuming there is a unitary operation U_b acting on $(l + 1)$ qubits which maps $|i\rangle |p\rangle$ to $|i\rangle |p \oplus b_i\rangle$ (see **Aside**),

---

Aside: U_b is unitary, that is, $U_b^\dagger U_b = I$ as we show now. Note that the states $|i\rangle |p\rangle$ for $i \in \{0, 1\}^l$ and $b \in \{0, 1\}$ form a basis, and let $|\psi\rangle = \sum_{i \in \{0,1\}^l, p \in \{0,1\}} \alpha_{i,p} |i\rangle |p\rangle$, then:

$\langle \psi | U_b^\dagger U_b |\psi\rangle$

$= (U_b |\psi\rangle)^\dagger U_b |\psi\rangle$

$= \left(U_b \left(\sum_{i \in \{0,1\}^l, p \in \{0,1\}} \alpha_{i,p} |i\rangle |p\rangle \right) \right)^\dagger U_b \left(\sum_{i \in \{0,1\}^l, p \in \{0,1\}} \alpha_{i,p} |i\rangle |p\rangle \right)$

$= \left(\sum_{i \in \{0,1\}^l, p \in \{0,1\}} \alpha_{i,p} U_b |i\rangle |p\rangle \right)^\dagger \left(\sum_{i \in \{0,1\}^l, p \in \{0,1\}} \alpha_{i,p} U_b |i\rangle |p\rangle \right)$

$= \left(\sum_{i \in \{0,1\}^l, p \in \{0,1\}} \alpha_{i,p} |i\rangle |p \oplus b_i\rangle \right)^\dagger \left(\sum_{i \in \{0,1\}^l, p \in \{0,1\}} \alpha_{i,p} |i\rangle |p \oplus b_i\rangle \right)$

$= \left(\sum_{i \in \{0,1\}^l, p \in \{0,1\}} \alpha_{i,p} \langle i| \langle p \oplus b_i| \right) \left(\sum_{i \in \{0,1\}^l, p \in \{0,1\}} \alpha_{i,p} |i\rangle |p \oplus b_i\rangle \right)$

$= \sum_{i \in \{0,1\}^l, p \in \{0,1\}} \alpha_{i,p}^2 \langle i| \langle p \oplus b_i| |i\rangle |p \oplus b_i\rangle$

$= \sum_{i \in \{0,1\}^l, p \in \{0,1\}} \alpha_{i,p}^2 = \langle \psi | \psi \rangle$

then apply U_b to $\frac{1}{\sqrt{n}} \sum_{i=0}^{n-1} |i\rangle \, |0\rangle$ (i.e. the result from (i) plus an ancillary qubit in state $|0\rangle$) to get:

$$U_b \left(\frac{1}{\sqrt{n}} \sum_{i=0}^{n-1} |i\rangle \, |0\rangle \right) = \frac{1}{\sqrt{n}} \sum_{i=0}^{n-1} U_b \, |i\rangle \, |0\rangle$$

(by linearity)

$$= \frac{1}{\sqrt{n}} \sum_{i=0}^{n-1} |i\rangle \, |0 \oplus b_i\rangle = \frac{1}{\sqrt{n}} \sum_{i=0}^{n-1} |i\rangle \, |b_i\rangle$$

(iii) and apply $I^{\otimes l} \otimes Z$ to the result from (ii) to get:

$$(I^{\otimes l} \otimes Z) \left(\frac{1}{\sqrt{n}} \sum_{i=0}^{n-1} |i\rangle \, |b_i\rangle \right) = \frac{1}{\sqrt{n}} \sum_{i=0}^{n-1} (-1)^{b_i} \, |i\rangle \, |b_i\rangle$$

(iv) and finally to "disentangle" the first l qubits from the last qubit, we apply U_b (again!) to the result from (iii) to get:

$$U_b \left(\frac{1}{\sqrt{n}} \sum_{i=0}^{n-1} (-1)_i^b \, |i\rangle \, |b_i\rangle \right)$$

$$= \frac{1}{\sqrt{n}} \sum_{i=0}^{n-1} (-1)^{b_i} U_b \, |i\rangle \, |b_i\rangle = \frac{1}{\sqrt{n}} \sum_{i=0}^{n-1} (-1)^{b_i} \, |i\rangle \, |b_i \oplus b_i\rangle$$

$$= \frac{1}{\sqrt{n}} \sum_{i=0}^{n-1} (-1)^{b_i} \, |i\rangle \, |0\rangle = \left(\frac{1}{\sqrt{n}} \sum_{i=0}^{n-1} (-1)^{b_i} \, |i\rangle \right) \otimes |0\rangle$$

That is, in (ii)–(iv), we applied $U_b(I^{\otimes l} \otimes Z)U_b$ and the last qubit can be measured (discarded) to get the state we wanted to prepare $\frac{1}{\sqrt{n}} \sum_{i=0}^{n-1} (-1)^{b_i} \, |i\rangle$; this l-qubit state is then sent to A (e.g. via teleportation – say, by teleporting each qubit separately).

2. A receives the state and applies an operation that maps $|i\rangle$ to $(-1)^{a_i} \, |i\rangle$, and then also applies the Hadamard transform to each qubit and measures the resulting l-qubit string.

The mapping from $|i\rangle$ to $(-1)^{a_i} \, |i\rangle$ can be done using an operation U_a defined similarly to U_b above, that is, we apply $U_a(I^{\otimes l} \otimes Z)U_a$ to the received state with an ancilla qubit $|0\rangle$:

$$U_a(I^{\otimes l} \otimes Z)U_a \left(\frac{1}{\sqrt{n}} \sum_{i=0}^{n-1} (-1)^{b_i} \, |i\rangle \right) |0\rangle = \left(\frac{1}{\sqrt{n}} \sum_{i=0}^{n-1} (-1)^{a_i + b_i} \, |i\rangle \right) |0\rangle$$

and the last qubit discarded. Then, after the Hadamard transforms, what is measured is the state:

$$H^{\otimes l} \left(\frac{1}{\sqrt{n}} \sum_{i=0}^{n-1} (-1)^{a_i + b_i} \, |i\rangle \right) = \frac{1}{\sqrt{n}} \sum_{i=0}^{n-1} (-1)^{a_i + b_i} H^{\otimes l} \, |i\rangle$$

$$= \frac{1}{\sqrt{n}} \sum_{i=0}^{n-1} (-1)^{a_i+b_i} (H \, |i_1\rangle \otimes \cdots \otimes H \, |i_l\rangle)$$

(where $|i\rangle = |i_1 \ldots i_l\rangle$, the bit string representation of i)

$$= \frac{1}{\sqrt{n}} \sum_{i=0}^{n-1} (-1)^{a_i+b_i} \left(\frac{1}{\sqrt{2}}\right)^l (|0\rangle + (-1)^{i_1} |1\rangle) \otimes \cdots \otimes (|0\rangle + (-1)^{i_l} |1\rangle)$$

$$= \frac{1}{n} \sum_{i=0}^{n-1} (-1)^{a_i+b_i} \sum_{j\in\{0,1\}^l} (-1)^{i\cdot j} |j\rangle$$

$$\left(\text{since } \left(\frac{1}{\sqrt{2}}\right)^l = \frac{1}{\sqrt{n}} \text{ and } i \cdot j = i_1 \times j_1 \oplus \cdots \oplus i_l \times j_l\right)$$

We note that if $a = b$, then the amplitude of $|\overbrace{0 \ldots 0}^{l}\rangle$ is $\frac{1}{n} \sum_{i=0}^{n-1} (-1)^{a_i+b_i} = 1$, i.e. if $a = b$, the probability of getting $|\overbrace{0 \ldots 0}^{l}\rangle$ is 1. But if a and b differs in exactly $n/2$ positions, then the amplitude $\frac{1}{n} \sum_{i=0}^{n-1} (-1)^{a_i+b_i} = 0$ (sum of an equal number of "1"s and "−1"s), and there is zero probability of getting the state $|\overbrace{0 \ldots 0}^{l}\rangle$.

3. A outputs 1 (i.e. we have case (1) $a = b$) if the measurement gave the state $|\overbrace{0 \ldots 0}^{l}\rangle$ or outputs 0 otherwise (i.e. we have case (2) a and b differ in exactly $n/2$ positions).

Hence, A is able to solve the promise problem with only $\log_2 n$ qubits sent from B! (As two classical bits (and one shared ebit) are required for teleporting each qubit, we would have $2(\log_2 n)$ or $O(\log n)$ classical bits required.) It was noted in Buhrman et al. [2010] that classical protocols need to send at least $0.007n$ bits! So, for small n, the classical protocols might be more efficient, but for larger n, the communication complexity for the classical protocols will be exponentially higher (but noting that the quantum protocol needs to use qubits).

3.1.1.3 The Distributed Intersection Problem

Let us consider another problem from Buhrman et al. [2010]. Suppose we have A and B, and A has an n-bit string $a = a_1 \ldots a_n$ and B has an n-bit string $b = b_1 \ldots b_n$. The problem is to compute the following function g:

$$g(a, b) = \begin{cases} 1, & \text{if } a_i = b_i = 1 \text{ for at least one } i \\ 0, & \text{otherwise} \end{cases}$$

One can imagine this as useful for computing a meeting time, where the bits of a and b represent slots when A and B are available ("0" for not available and "1" for available) and trying to find at least one slot where both are available.

This problem is equivalent to finding indices i such that $a_i \wedge b_i = 1$. For example, if $a = 10111$ and $b = 11101$, then $a_i \wedge b_i = 1$ when i is 0, 2 and 4. Hence, given two n-bit strings a and b, this becomes a search problem, that of finding indices $i \in \{1, \ldots, n\}$ satisfying the criteria $a_i \wedge b_i = 1$ (we could also use indices 0 to $n-1$).

We can use Grover's (quantum) algorithm [Grover, 1996] to perform this search, which allows us to search for such a solution (asymptotically) faster than the best classical algorithm. Grover's algorithm uses a function $f : \{0, \ldots, n-1\} \to \{0,1\}$ such that $f(x) = 1$ if and only if x points to the item that satisfies the search criterion – we can think of x as an index into a database of items – let us just call f a database. Suppose x' is an item such that $f(x') = 1$, we assume we have a quantum operation $V_{x'}$ that can use f defined as follows[2]:

$$V_{x'} |x\rangle = \begin{cases} -|x\rangle, & \text{if } x = x', \text{ i.e. } f(x) = 1 \\ |x\rangle, & \text{if } x \neq x', \text{ i.e. } f(x) = 0 \end{cases}$$

that is, $V_{x'}$ can be written as $V_{x'} |x\rangle = (-1)^{f(x)} |x\rangle$. The idea of the operation $V_{x'}$ is to "mark" with a "-1" the item that satisfies a particular search criterion. In Grover's algorithm, there is a unitary operation called the *Grover iterate* defined as $G = H^{\otimes l} V_0 H^{\otimes l} V_{x'}$, assuming $n = 2^l$. The Grover iterate is repeatedly applied, first to a state $|s\rangle$ as below, and then repeatedly to the resulting state:

1. from the l-qubit state $|0\rangle^{\otimes l}$, prepare the state $|s\rangle$, e.g. via $|s\rangle = H^{\otimes l} |0\rangle^{\otimes l} = \frac{1}{\sqrt{n}} \sum_{i=0}^{n-1} |i\rangle$
2. apply the Grover iterate G, t times to the state
3. measure all the qubits in the resulting state in the computational basis

At the end of the above, one can check if the result obtained (the string of bits obtained from (3)) is indeed a solution to the problem. Grover's algorithm can return $|x'\rangle$ with probability approaching 1 using $O(\sqrt{n})$ applications of $V_{x'}$, in particular when $t = \lceil \frac{\pi}{4}(\sqrt{n}) \rceil$. If there are multiple solutions, say m solutions, then $t = \lceil \frac{\pi}{4}(\sqrt{n/m}) \rceil$. (Note that if after running $\lceil \frac{\pi}{4}(\sqrt{n}) \rceil$ iterations and the result obtained is not a solution, likely, there was likely no solution to start with!)

We will look at how the intersection problem can be an application of Grover's algorithm. If a and b are on the same machine, we can define f such that $f(x) = 1$ if and only if $a_x \wedge b_x = 1$, i.e. $f(x) = a_x \wedge b_x$. Then, the operation $V_{a,b}$ that depends on a and b then becomes:

$$V_{a,b} |x\rangle = (-1)^{a_x \wedge b_x} |x\rangle$$

---

2 Alternatively, we could define this as $V_{x'} = I - 2 |x'\rangle \langle x'|$, with an $l \times l$ identity matrix I.

Using this operation repeatedly in the Grover iterate with $O(\sqrt{n})$ then yields a solution to the problem.

But our problem is distributed with A having a and B having b. So, how can A run Grover's algorithm but perform the above $V_{a,b}$ operations in a distributed way? One way to do this is as follows.

Algorithm to Implement $V_{a,b}$

Whenever A wants to perform $V_{a,b}$ on some state $|\psi\rangle = \sum_{i=0}^{n-1} \alpha_i |i\rangle$, that is, A wants to obtain

$$V_{a,b} |\psi\rangle = V_{a,b} \left(\sum_{i=0}^{n-1} \alpha_i |i\rangle \right) = \sum_{i=0}^{n-1} \alpha_i V_{a,b} |i\rangle = \sum_{i=0}^{n-1} \alpha_i (-1)^{a_i \wedge b_i} |i\rangle$$

machine A can do the following:

1. Machine A creates the state

$$\sum_{i=0}^{n-1} \alpha_i |i\rangle |a_i\rangle$$

   effectively tagging on a_i as an additional qubit to the state $|\psi\rangle$. This state can be prepared as follows: suppose we have an operation $U_a |i\rangle |p\rangle = |i\rangle |p \oplus a_i\rangle$, which we have shown to be a unitary operation previously, then apply U_a to $|\psi\rangle |0\rangle$, that is, $U_a |\psi\rangle |0\rangle = U_a(\sum_{i=0}^{n-1} \alpha_i |i\rangle) |0\rangle = \sum_{i=0}^{n-1} \alpha_i U_a |i\rangle |0\rangle = \sum_{i=0}^{n-1} \alpha_i |i\rangle |0 \oplus a_i\rangle = \sum_{i=0}^{n-1} \alpha_i |i\rangle |a_i\rangle$.

2. Then, A sends B the state $\sum_{i=0}^{n-1} \alpha_i |i\rangle |a_i\rangle$, which is $(l+1)$ qubits (since we are assuming $n = 2^l$, for simplicity).

3. Machine B then applies the unitary operation: $O_b : |i\rangle |x\rangle \rightarrow (-1)^{x \wedge b_i} |i\rangle |x\rangle$ on the received state $\sum_{i=0}^{n-1} \alpha_i |i\rangle |a_i\rangle$ to obtain:

$$O_b \left(\sum_{i=0}^{n-1} \alpha_i |i\rangle |a_i\rangle \right) = \sum_{i=0}^{n-1} \alpha_i O_b |i\rangle |a_i\rangle = \sum_{i=0}^{n-1} \alpha_i (-1)^{a_i \wedge b_i} |i\rangle |a_i\rangle$$

   (Note that the operation O_b on some state $|i\rangle |x\rangle$ can be implemented as follows: assuming we have an operation U'_b that maps $|i\rangle |p\rangle |q\rangle$ to $|i\rangle |p\rangle |q \oplus (p \wedge b_i)\rangle$,[3]

---

3 This operation is unitary since

$\langle \psi | U'^{\dagger}_b U'_b |\psi\rangle$

$= (U'_b |\psi\rangle)^{\dagger} U'_b |\psi\rangle$

$= \left(U'_b \left(\sum_{i \in \{0,1\}^l, p \in \{0,1\}, q \in \{0,1\}} \alpha_{i,p,q} |i\rangle |p\rangle |q\rangle \right) \right)^{\dagger} U'_b \left(\sum_{i \in \{0,1\}^l, p \in \{0,1\}, q \in \{0,1\}} \alpha_{i,p,q} |i\rangle |p\rangle |q\rangle \right)$

$= \left(\sum_{i \in \{0,1\}^l, p \in \{0,1\}, q \in \{0,1\}} \alpha_{i,p,q} U'_b |i\rangle |p\rangle |q\rangle \right)^{\dagger} \left(\sum_{i \in \{0,1\}^l, p \in \{0,1\}, q \in \{0,1\}} \alpha_{i,p,q} U'_b |i\rangle |p\rangle |q\rangle \right)$

perform the computation

$$U_b'(I^{\otimes(l+1)} \otimes Z)U_b' |i\rangle |x\rangle |0\rangle$$
$$= U_b'(I^{\otimes(l+1)} \otimes Z) |i\rangle |x\rangle |0 \oplus (x \wedge b_i)\rangle = U_b'(-1)^{x \wedge b_i} |i\rangle |x\rangle |x \wedge b_i\rangle$$
$$= (-1)^{x \wedge b_i} U_b' |i\rangle |x\rangle |x \wedge b_i\rangle = (-1)^{x \wedge b_i} |i\rangle |x\rangle |(x \wedge b_i) \oplus (x \wedge b_i)\rangle$$
$$= (-1)^{x \wedge b_i} |i\rangle |x\rangle |0\rangle$$

and discarding the ancilla qubit $|0\rangle$, we have mapped $|i\rangle |x\rangle$ to $(-1)^{x \wedge b_i} |i\rangle |x\rangle$.)

4. Machine B then sends the result $\sum_{i=0}^{n-1} \alpha_i(-1)^{a_i \wedge b_i} |i\rangle |a_i\rangle$ back to A, i.e. $(l+1)$ qubits.

5. Machine A receives the result and sets the last qubit to $|0\rangle$ by applying U_a:

$$U_a \left(\sum_{i=0}^{n-1} \alpha_i(-1)^{a_i \wedge b_i} |i\rangle |a_i\rangle \right) = \sum_{i=0}^{n-1} \alpha_i(-1)^{a_i \wedge b_i} U_a |i\rangle |a_i\rangle$$
$$= \sum_{i=0}^{n-1} \alpha_i(-1)^{a_i \wedge b_i} |i\rangle |a_i \oplus a_i\rangle = \sum_{i=0}^{n-1} \alpha_i(-1)^{a_i \wedge b_i} |i\rangle |0\rangle$$

and the last qubit can be discarded/ignored, and so machine A obtains:

$$\sum_{i=0}^{n-1} \alpha_i(-1)^{a_i \wedge b_i} |i\rangle$$

as required, and we have computed $V_{a,b} |\psi\rangle$!

In summary, the algorithm to compute a solution to the distributed intersection problem is as follows, perform at A:

1. from the l-qubit state $|0\rangle^{\otimes l}$, prepare the state $|s\rangle = \frac{1}{\sqrt{n}} \sum_{i=0}^{n-1} |i\rangle$, e.g. via $|s\rangle = H^{\otimes l} |0\rangle^{\otimes l}$

2. apply the Grover iterate G, t times, starting with the state $|s\rangle$, where G is defined as follows (equivalent to the previous form): $G = (2 |s\rangle \langle s| - I)V_{a,b}$, where each time $V_{a,b}$ is applied, communication with B is involved (as explained above); that is, we compute $\underbrace{G \ldots G}_{t} |s\rangle$

$$= \left(\sum_{i \in \{0,1\}^l, p \in \{0,1\}, q \in \{0,1\}} \alpha_{i,p,q} |i\rangle |p\rangle |q \oplus (p \wedge b_i)\rangle \right)^{\dagger} \left(\sum_{i \in \{0,1\}^l, p \in \{0,1\}, q \in \{0,1\}} \alpha_{i,p,q} |i\rangle |p\rangle |q \oplus (p \wedge b_i)\rangle \right)$$

$$= \left(\sum_{i \in \{0,1\}^l, p \in \{0,1\}, q \in \{0,1\}} \alpha_{i,p,q} \langle i| \langle p| \langle q \oplus (p \wedge b_i)| \right) \left(\sum_{i \in \{0,1\}^l, p \in \{0,1\}, q \in \{0,1\}} \alpha_{i,p,q} |i\rangle |p\rangle |q \oplus (p \wedge b_i)\rangle \right)$$

$$= \sum_{i \in \{0,1\}^l, p \in \{0,1\}, q \in \{0,1\}} \alpha_{i,p,q}^2 \langle i| \langle p| \langle q \oplus (p \wedge b_i)| |i\rangle |p\rangle |q \oplus (p \wedge b_i)\rangle$$

$$= \sum_{i \in \{0,1\}^l, p \in \{0,1\}, q \in \{0,1\}} \alpha_{i,p,q}^2 = \langle \psi| \psi \rangle$$

3. measure all the qubits in the resulting state in the computational basis

Analysis: In each Grover iterate operation involving $V_{a,b}$, two messages, each of $(l + 1) = (log_2 \; n + 1)$ qubits, are sent, and there would be $t = O(\sqrt{n})$ iterations; so, the algorithm uses a total of $O(\sqrt{n} \cdot log \; n)$ qubits of communication. Each qubit, sent by teleportation, would use a shared ebit and two classical bits, and so, $O(\sqrt{n} \cdot log \; n)$ classical bits of communication is required. In fact, a more efficient algorithm can do better – communicating only $O(\sqrt{n})$ qubits [Buhrman et al., 2010]. This contrasts with the classical algorithm which would require communicating about n bits, and so we have a quadratic gap of improvement in communication complexity with the quantum solution (particularly notable when n is very large)!

3.1.1.4 Discussion

We have seen an example of a distributed computation where the communication complexity was reduced using entanglement, by one bit, compared to the best possible classical protocol (without entanglement), an example on computing the DJ Promise Problem where the communication complexity was reduced by an exponential gap by transferring quantum states (e.g. using teleportation which might use entanglement), and a third example on computing the distributed intersection where a quadratic improvement gap is obtained. This leads to the following question raised by Buhrman et al. [2010]: "Could quantum communication complexity be much more efficient for every communication problem?" Unfortunately, the answer to the question is in the negative, and it was noted that for most communication complexity problems, "quantum communication does not help much"; for example, computing the inner product (of two n-bit strings with A having one and B having the other) was pointed out as an example where the best classical and quantum protocols both need n qubits/bits to be communicated to solve.

3.1.2 Enabling Distributed Computations Not Possible Classically

We have looked at how certain distributed computation problems can be solved with lower communication complexity with the use of quantum communication. We now look at certain distributed computation problems which cannot be solved without the use of quantum protocols, in particular, without entanglement. We look at two problems from Buhrman et al. [2010].

3.1.2.1 Greenberger–Horne–Zeilinger and Mermin (GHZ&M) Game

The name GHZ&M comes from the earlier authors on this problem [Buhrman et al., 2010]. Consider three parties with the inputs and outputs shown in

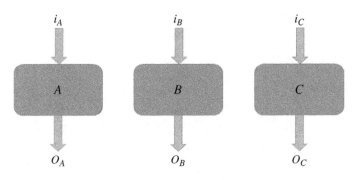

Figure 3.1   The GHZ&M problem is for A, B, and C, on receiving inputs i_A, i_B, and i_C respectively, to compute the output bits o_A, o_B, and o_C (respectively), such that certain properties among the output bits are satisfied, without A, B, and C communicating once they receive their inputs.

Figure 3.1. The GHZ&M problem is to compute the output bits o_A, o_B, and o_C such that the following condition is satisfied:

$$o_A \oplus o_B \oplus o_C = \begin{cases} 0, & \text{if } i_A i_B i_C = 000 \\ 1, & \text{if } i_A i_B i_C \in \{011, 101, 110\} \end{cases}$$

but no communication (sending of classical or quantum bits) is allowed once A, B, and C receive their inputs. Note that if the inputs are anything else other than the four combinations above, it doesn't matter what the output is (or we assume that the inputs are always one of the four combinations $\{000, 011, 101, 110\}$, each occurring with equal probability).

Let's look at the quantum solution. Suppose A, B, and C each has one qubit ψ_A, ψ_B and ψ_C, respectively, and the state of the 3-qubit system is the entangled state:

$$\frac{1}{2}(|000\rangle - |011\rangle - |101\rangle - |110\rangle)$$

which is equivalent to the GHZ state under LU operations – see **Aside**.

Aside: There could be different possible operations; one is as follows:

1. starting with $\frac{1}{\sqrt{2}}(|000\rangle + |111\rangle)$, we apply the operation $H \otimes H \otimes H$:

$$(H \otimes H \otimes H)\,\frac{1}{\sqrt{2}}(|000\rangle + |111\rangle)$$

$$= \frac{1}{\sqrt{2}}(H|0\rangle\,H|0\rangle\,H|0\rangle + H|1\rangle\,H|1\rangle\,H|1\rangle)$$

$$= \frac{1}{4}((|000\rangle + |001\rangle + |010\rangle + |011\rangle + |100\rangle + |101\rangle + |110\rangle + |111\rangle)$$

$$+ (|000\rangle - |001\rangle - |010\rangle + |011\rangle - |100\rangle + |101\rangle + |110\rangle - |111\rangle)))$$

$$= \frac{1}{2}(|000\rangle + |011\rangle + |101\rangle + |110\rangle)$$

2. apply the operation $\sqrt{Z} \otimes \sqrt{Z} \otimes \sqrt{Z}$ (and recalling that $\sqrt{Z}|0\rangle = |0\rangle$ and $\sqrt{Z}|1\rangle = i|1\rangle$): $(\sqrt{Z} \otimes \sqrt{Z} \otimes \sqrt{Z}) \frac{1}{2}(|000\rangle + |011\rangle + |101\rangle + |110\rangle) = \frac{1}{2}(|000\rangle + i^2 |011\rangle + i^2 |101\rangle + i^2 |110\rangle)$, and since $i^2 = -1$, we get $\frac{1}{2}(|000\rangle - |011\rangle - |101\rangle - |110\rangle)$ as required.

The procedure for A, B, and C is as follows:

0.	**for each** $p \in \{A, B, C\}$:
1.	**if** $i_p = 1$ **then**
2.	**apply** H **to** ψ_p
3.	**else**
4.	**do nothing to** ψ_p
5.	**measure** ψ_p **in the computational basis yielding bit** o_p
6.	**compute** $o_A \oplus o_B \oplus o_C$

We can show that this works – note that no communication among A, B, and C is involved in the above procedure.

When $i_A i_B i_C = 000$, the states are measured in the computational basis without applying any H gate, and so, we get as output bits one of $\{|000\rangle, |011\rangle, |101\rangle, |110\rangle\}$, which is in fact, by choice of our entangled 3-qubit system state we wanted A, B, and C to share (each possibility with an even number of ones), so that the $o_A \oplus o_B \oplus o_C = 0$, as required.

When $i_A i_B i_C \in \{011, 101, 110\}$, we apply H to exactly two of the three qubits with the effect of generating a superposition of states with an odd number of ones, e.g. when $i_A i_B i_C = 011$:

$$(I \otimes H \otimes H)\frac{1}{2}(|000\rangle - |011\rangle - |101\rangle - |110\rangle)$$

$$= \frac{1}{2}(|0\rangle (H|0\rangle H|0\rangle - H|1\rangle H|1\rangle) - |1\rangle (H|0\rangle H|1\rangle + H|1\rangle H|0\rangle))$$

$$= \frac{1}{2}(|0\rangle (|01\rangle + |10\rangle) - |1\rangle (|00\rangle - |11\rangle))$$

$$= \frac{1}{2}(|001\rangle + |010\rangle - |100\rangle + |111\rangle)$$

And since there will be an odd number of ones obtained when the qubits are measured, whether we get 001, 010, 100, or 111, we have $o_A \oplus o_B \oplus o_C = 1$, as required; similarly, with $i_A i_B i_C = 101$ and $i_A i_B i_C = 110$.

Hence, without any communication, but with the shared 3-qubit entangled state, we can solve the GHZ&M problem. We will see that there this problem cannot be solved in the classical case, without communication.

In the classical case, without any shared entangled state, each party can set its strategy beforehand to output either 0 or 1 when the input is 0, and to output 0 or 1 when the input is 1. Let us say A has the following strategy A_0 and A_1, representing A's output when the input is 0 and 1, respectively. Similarly for B and C, i.e. B_0 and B_1 and C_0 and C_1, respectively. Then, the condition on the outputs can be equivalently represented by this set of simultaneous equations:

$$A_0 \oplus B_0 \oplus C_0 = 0; \quad A_0 \oplus B_1 \oplus C_1 = 1; \quad A_1 \oplus B_0 \oplus C_1 = 1;$$
$$A_1 \oplus B_1 \oplus C_0 = 1$$

Can these equations be satisfied simultaneously? No! We can see this by summing (modulo 2) the LHS (left hand side) of the four equations, we get 0, and summing (modulo 2) the RHS of the four equations, we get 1, and $0 \neq 1$! This means that there is no set of strategies that the parties can set to satisfy the condition on the outputs.

Suppose $i_A i_B i_C$ can take any of the combinations in $\{000, 011, 101, 110\}$ with equal probability. What is the best probability of success in the classical case? Note that any three of the above four equations can be satisfied simultaneously, that is, there is a strategy for each of A, B, C, which when adopted will satisfy three of the four cases. For example, suppose we choose the first three equations – we can solve these three simultaneously to get a strategy for A, B, and C, which is one satisfying:

$$(A_0 \oplus B_0 \oplus C_0) \oplus (A_0 \oplus B_1 \oplus C_1) \oplus (A_1 \oplus B_0 \oplus C_1) = A_1 \oplus B_1 \oplus C_0 = 0$$

Suppose we set $A_1 = B_1 = C_0 = 0$ satisfying the above equation, and using these values and the first equation, set $A_0 = B_0 = 0$ and then using the values we have and the second equation set $C_1 = 1$; we then have a strategy for A, B, and C satisfying the first three equations and will work whenever the input is one of $\{000, 011, 101\}$, that is, since each combination can appear with equal probability, this strategy will work three times out of four, or with probability 3/4. The same argument goes when any three of the four equations are selected, instead of the first three. For any three of the four equations selected, we will have a strategy for A, B, and C, and this strategy will have success probability 3/4, and there are only four ways to choose three out of four equations ($^4C_3$), and so, there are four possible strategies to choose from. One can have the three parties randomly select any one of the four strategies before receiving the inputs, so that the probability of success is $1/4 \cdot 3/4 + 1/4 \cdot 3/4 + 1/4 \cdot 3/4 + 1/4 \cdot 3/4 = 3/4$, where the product $1/4 \cdot 3/4$ is from the probability of choosing a strategy (1/4) and the probability of winning with that chosen strategy (3/4), and summing over four possible strategies. Hence, it is not possible to have a success probability better than 3/4.

The quantum approach, therefore, enables a distributed computation not possible to be done classically with probability $>3/4$!

3.1.2.2 Clauser–Horne–Shimony–Holt (CHSH)

Let us look at another problem (or one could call this a game!) involving two parties A and B, in each round of the game, each receiving one input bit, i_A and i_B, respectively, and the problem is for them to compute output bits o_A and o_B, respectively, to satisfy the following condition:

$$o_A \oplus o_B = i_A \wedge i_B$$

as illustrated in Figure 3.2, without communicating when they received their inputs (so, A does not know B's inputs and conversely). In a round, A and B *win* provided their outputs are such that the above condition is satisfied and lose otherwise. If A and B cannot always win, then the aim is for A and B to have some algorithm/strategy that will maximize the probability of winning.

Let us look at a quantum solution for this problem. Assume that A and B have two qubits $\psi_A \psi_B$, one qubit each, in the following shared entangled state between them: $\frac{1}{\sqrt{2}}(|00\rangle - |11\rangle)$. Note that this state can be obtained from the state $\frac{1}{\sqrt{2}}(|00\rangle + |11\rangle)$ by applying $I \otimes Z$.

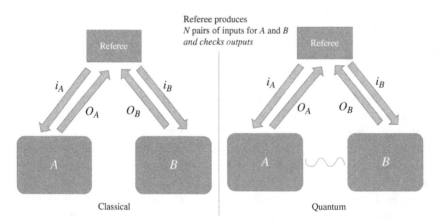

Figure 3.2 The CHSH game. The referee produces a pair of inputs i_A and i_B for A and B in each round, say for N rounds. In each round, A outputs o_A and B outputs o_B in response to the inputs and cannot communicate on receiving the inputs. In the quantum case, the boxes share an entangled state, a "fresh" one for each round. Upon receiving the outputs, the referee checks for the winning condition $o_A \oplus o_B = i_A \wedge i_B$. A question is: what can A and B do, in the classical (and in the quantum case), so that out of N rounds, they win as many rounds as possible? And can the quantum case do better (have more wins, on average) than the classical case?

Now let the unitary operation that rotates a qubit by angle θ be denoted by

$$R(\theta) = \begin{bmatrix} \cos\theta & -\sin\theta \\ \sin\theta & \cos\theta \end{bmatrix}$$

The algorithm for both A and B (let us call this the *CHSH algorithm*, for simplicity) that will compute outputs that will maximize the probability that the condition is satisfied is as follows:

For A:

1.   **if $i_A = 0$ then**
2.   **apply $R(-\pi/16)$ to ψ_A**
3.   **else (if $i_A = 1$)**
4.   **apply $R(3\pi/16)$ to ψ_A**
5.   **measure ψ_A in the computational basis yielding bit o_A**

---

and for B, it is practically the same – just replace A by B in the above.

Observe that $R(\theta)|0\rangle = \cos\theta\,|0\rangle + \sin\theta\,|1\rangle$ and $R(\theta)|1\rangle = -\sin\theta\,|0\rangle + \cos\theta\,|1\rangle$. Then, if A rotates its qubit by θ_1 and B rotates its qubit by θ_2, the entangled state becomes:

$$(R(\theta_1) \otimes R(\theta_2)) \frac{1}{\sqrt{2}}(|00\rangle - |11\rangle)$$

$$= \frac{1}{\sqrt{2}}(R(\theta_1)|0\rangle \otimes R(\theta_2)|0\rangle - R(\theta_1)|1\rangle \otimes R(\theta_2)|1\rangle)$$

$$= \frac{1}{\sqrt{2}}((\cos\theta_1|0\rangle + \sin\theta_1|1\rangle) \otimes (\cos\theta_2|0\rangle + \sin\theta_2|1\rangle)$$

$$- (-\sin\theta_1|0\rangle + \cos\theta_1|1\rangle) \otimes (-\sin\theta_2|0\rangle + \cos\theta_2|1\rangle))$$

$$= \frac{1}{\sqrt{2}}((\cos\theta_1\cos\theta_2|00\rangle + \cos\theta_1\sin\theta_2|01\rangle + \sin\theta_1\cos\theta_2|10\rangle + \sin\theta_1\sin\theta_2|11\rangle)$$

$$- (\sin\theta_1\sin\theta_2|00\rangle - \sin\theta_1\cos\theta_2|01\rangle - \cos\theta_1\sin\theta_2|10\rangle + \cos\theta_1\cos\theta_2|11\rangle))$$

$$= \frac{1}{\sqrt{2}}((\cos\theta_1\cos\theta_2 - \sin\theta_1\sin\theta_2)|00\rangle + (\cos\theta_1\sin\theta_2 + \sin\theta_1\cos\theta_2)|01\rangle$$

$$+ (\sin\theta_1\cos\theta_2 + \cos\theta_1\sin\theta_2)|10\rangle + (\sin\theta_1\sin\theta_2 - \cos\theta_1\cos\theta_2)|11\rangle)$$

$$= \frac{1}{\sqrt{2}}(\cos(\theta_1 + \theta_2)|00\rangle + \sin(\theta_1 + \theta_2)|01\rangle + \sin(\theta_1 + \theta_2)|10\rangle - \cos(\theta_1 + \theta_2)|11\rangle)$$

(since $\cos(\theta_1 + \theta_2) = \cos\theta_1\cos\theta_2 - \sin\theta_1\sin\theta_2$ and

$\sin(\theta_1 + \theta_2) = \cos\theta_1\sin\theta_2 + \sin\theta_1\cos\theta_2$)

Then, after the measurements, if we obtain $|00\rangle$ or $|11\rangle$, then $o_A = o_B$, i.e. $o_A \oplus o_B = 0$, and this occurs with probability $\frac{1}{2}\cos^2(\theta_1 + \theta_2) + \frac{1}{2}\cos^2(\theta_1 + \theta_2) = \cos^2(\theta_1 + \theta_2)$ and if we obtain $|01\rangle$ or $|10\rangle$, then $o_A \neq o_B$, i.e. $o_A \oplus o_B = 1$, and this occurs with probability $\frac{1}{2}\sin^2(\theta_1 + \theta_2) + \frac{1}{2}\sin^2(\theta_1 + \theta_2) = \sin^2(\theta_1 + \theta_2)$.

We can compute the probability that the required condition for this problem, which is $o_A \oplus o_B = i_A \wedge i_B$, is satisfied, as follows. The following table considers all the input possibilities, the rotation angle to be applied to each qubit in the shared ebit according to the procedure for A and B above, and the probability $prob(o_A \oplus o_B = i_A \wedge i_B)$.

i_A	i_B	$R(\theta_1)$	$R(\theta_2)$	$i_A \wedge i_B$	$prob(o_A \oplus o_B = i_A \wedge i_B)$
0	0	$R(-\pi/16)$	$R(-\pi/16)$	0	$prob(00 \text{ or } 11) = \cos^2((-\pi/16) + (-\pi/16))$
0	1	$R(-\pi/16)$	$R(3\pi/16)$	0	$prob(00 \text{ or } 11) = \cos^2((-\pi/16) + (3\pi/16))$
1	0	$R(3\pi/16)$	$R(-\pi/16)$	0	$prob(00 \text{ or } 11) = \cos^2((3\pi/16) + (-\pi/16))$
1	1	$R(3\pi/16)$	$R(3\pi/16)$	1	$prob(01 \text{ or } 10) = \sin^2((3\pi/16) + (3\pi/16))$

We can see from the table, that in all four input possibilities, $prob(o_A \oplus o_B = i_A \wedge i_B)$ is $\cos^2(\pi/8) \approx 0.8536$ (in the last case, note that $\sin^2((3\pi/16) + (3\pi/16)) = \sin^2(3\pi/8) = \cos^2(\pi/8)$). Hence, with entanglement, this problem can be solved with probability $\approx 0.8536 > 0.75$ by the procedure above, where 0.75 is the best classical solution, as we show below. With such probabilities, with uniformly random inputs (over 00, 01, 10, and 11), in one round, the quantum case has a better chance of winning, and after N rounds, for sufficiently large N, in the quantum case, we will see that $\approx 0.8536N$ of the rounds can be a win whereas the classical case can expect at most around $0.75N$ wins.

In the classical case, let us say A has the following strategy A_0 and A_1, representing A's output when the input is 0 and 1, respectively, and similarly with B. Then, the condition on the outputs can be represented by this set of simultaneous equations:

$$A_0 \oplus B_0 = 0; \quad A_0 \oplus B_1 = 0; \quad A_1 \oplus B_0 = 0; \quad A_1 \oplus B_1 = 1$$

We can see that it is not possible to have a solution that simultaneously satisfies all four equations since when we sum (modulo 2) the LHS and the RHS, they are not equal: $(A_0 \oplus B_0) \oplus (A_0 \oplus B_1) \oplus (A_1 \oplus B_0) \oplus (A_1 \oplus B_1) = 0 \neq 0 \oplus 0 \oplus 0 \oplus 1 = 1$. But by using a probabilistic strategy, a success probability of 3/4 is possible in the classical case – similar to the reasoning with GHZ&M we saw earlier. A and B can randomly select any three of the four equations to satisfy and come up with a strategy. For example, they select the first three equations and agree on a strategy consistent with satisfying the first three equations simultaneously (e.g. by setting $A_0 = A_1 = B_0 = B_1 = 1$), and then when the inputs are 00, 01, or 10, then the strategy will succeed (but when the inputs are 11, the condition will not be satisfied and the strategy will fail), and so, 3/4 of the time (if the input combinations occur with uniform probability). This is the same for any three equations selected to be satisfied, and so, this is the best success probability in the classical case is 3/4.

Hence, the quantum solution, with entanglement, wins with higher probability compared to the best classical case. Note that this is related to the CHSH inequality (which we explain further below) for "testing" for entangled Bell states. The idea is to play this game and then gather winning statistics and if the statistics exceeds 0.75 and approaches that in the quantum case of 0.8536, then there must have been entanglement used since a classical approach cannot do better than 0.75 – we will see that this is useful later for security testing in quantum key distribution (QKD).

Another "Game" but the Same Quantum Solution

Suppose we change the game slightly so that the set up is the same as shown in Figure 3.2 (i.e. again, A and B receive inputs, and on receiving inputs, they are not allowed to communicate), but now, the aim is for A and B to maximize the following quantity S (let us call it the CHSH expression) which is the absolute value of a sum of mean values, after playing N rounds:

$$S = |\langle A_0 B_0 \rangle + \langle A_0 B_1 \rangle + \langle A_1 B_0 \rangle - \langle A_1 B_1 \rangle|$$

where $\langle A_i B_j \rangle = \frac{1}{N_{ij}} \sum_{v=1}^{N_{ij}} A_i^v B_j^v$, and in round v, A_i has value $A_i^v \in \{+1, -1\}$ and B_j has value $B_j^v \in \{+1, -1\}$, and N_{ij} ($< N$, where we assume that each combination of inputs 00, 01, 10, and 11 occurs with equal probability in each round, so that we have $N_{ij} = \frac{N}{4}$, for each combination ij) is the number of rounds with input to A being i and input to B being j, i.e. $\langle A_i B_j \rangle$ is the average value of the product of A's output and B's output given the inputs i and j, respectively. Note in this case, the outputs are either $+1$ or -1, A_0 denotes the output of A in response to input 0, and similarly for, symbols A_1, B_0, and B_1.

Since the value of $A_0 B_0 + A_0 B_1 + A_1 B_0 - A_1 B_1 = 2$ or -2 (verified by substituting in all possible values to check – see **Aside**), we have $S \leq 2$. This expression is called the *CHSH inequality*, named after the discoverers (i.e. John Clauser, Michael Horne, Abner Shimony, and Richard Holt) of the idea (the first mentioned in Chapter 1 who won the Nobel Prize in Physics) that there is a bound on certain classical correlations, which a quantum approach can violate.

Aside: Possible values for the expression $A_0B_0 + A_0B_1 + A_1B_0 - A_1B_1$:

A_0	A_1	B_0	B_1	$A_0B_0 + A_0B_1 + A_1B_0 - A_1B_1$
+1	+1	+1	+1	2
+1	+1	+1	−1	2
+1	+1	−1	+1	−2
+1	+1	−1	−1	−2
+1	−1	+1	+1	2
+1	−1	+1	−1	−2
+1	−1	−1	+1	2
+1	−1	−1	−1	−2
−1	+1	+1	+1	−2
−1	+1	+1	−1	2
−1	+1	−1	+1	−2
−1	+1	−1	−1	2
−1	−1	+1	+1	−2
−1	−1	+1	−1	−2
−1	−1	−1	+1	2
−1	−1	−1	−1	2

For the quantum case, using the same CHSH algorithm (and the entangled state $\frac{1}{\sqrt{2}}(|00\rangle - |11\rangle)$ shared by A and B) that was used earlier to win the CHSH game with probability ≈ 0.8536, would it possible to get a higher value for S, that is, violate the CHSH inequality? The answer is yes! We explain how below.

In the quantum case, using the same algorithm above in A and B, after N rounds what would we expect the value of $\langle A_iB_j \rangle$ to be?

In general, according to the rules of quantum mechanics, the expected value (denoted by $E(\theta_1, \theta_2)$) of the product of A and B's outputs, $o_A o_B$, we get from the measurement after using $R(\theta_1)$ at A and $R(\theta_2)$ at B is

$$E(\theta_1, \theta_2) = \sum_{o_A, o_B \in \{+1, -1\}} o_A o_B \cdot P_{o_A, o_B}(\theta_1, \theta_2)$$

$$= (+1)(+1) \cdot P_{+1,+1}(\theta_1, \theta_2) + (-1)(-1) \cdot P_{-1,-1}(\theta_1, \theta_2)$$
$$+ (+1)(-1) \cdot P_{+1,-1}(\theta_1, \theta_2) + (-1)(+1) \cdot P_{-1,+1}(\theta_1, \theta_2)$$

where $P_{+1,+1}(\theta_1, \theta_2)$ denotes the probability of obtaining +1 and +1 (or the state $|00\rangle$, with eigenvalue +1) when measuring using the basis $B =$

$\{R(\theta_1) \otimes R(\theta_2) |00\rangle, R(\theta_1) \otimes R(\theta_2) |01\rangle, R(\theta_1) \otimes R(\theta_2) |10\rangle, R(\theta_1) \otimes R(\theta_2) |11\rangle\}$,[4] and similarly, for $P_{-1,-1}(\theta_1, \theta_2)$ (corresponding to the state $|11\rangle$, with eigenvalue +1), $P_{+1,-1}(\theta_1, \theta_2)$ (corresponding to the state $|01\rangle$, with eigenvalue −1), and $P_{-1,+1}(\theta_1, \theta_2)$ (corresponding to the state $|10\rangle$, with eigenvalue −1).

Recall that in the CHSH algorithm (at the beginning of Section 3.1.2.2), when the input to A is 0, the operator $R(-\pi/16)$ is applied and measurement is done to produce an output, and when the input to A is 1, the operator $R(3\pi/16)$ is applied and measurement is done to produce an output; similarly, with B. After measurement, now looking at the eigenvalue obtained (rather than the resulting quantum state), we get +1 or −1. In other words, with inputs $i_A = 0$, and $i_B = 0$, the algorithm applies $R(-\pi/16)$ at A and $R(-\pi/16)$ at B and we have:

$$E(-\pi/16, -\pi/16) = P_{+1,+1}(-\pi/16, -\pi/16) + P_{-1,-1}(-\pi/16, -\pi/16)$$
$$- P_{+1,-1}(-\pi/16, -\pi/16) - P_{-1,+1}(-\pi/16, -\pi/16)$$

And for large enough N, the average value of the product $A_0 B_0$

$$\langle A_0 B_0 \rangle \approx E(-\pi/16, -\pi/16)$$

that is, after N rounds, the statistics should show that the average $\langle A_0 B_0 \rangle$ is approximately $E(-\pi/16, -\pi/16)$. And similar reasoning leads to:

$$\langle A_0 B_1 \rangle \approx E(-\pi/16, 3\pi/16)$$
$$\langle A_1 B_0 \rangle \approx E(3\pi/16, -\pi/16)$$
$$\langle A_1 B_1 \rangle \approx E(3\pi/16, 3\pi/16)$$

This means that we can expect

$$S = |\langle A_0 B_0 \rangle + \langle A_0 B_1 \rangle + \langle A_1 B_0 \rangle - \langle A_1 B_1 \rangle|$$
$$\approx |E(-\pi/16, -\pi/16) + E(-\pi/16, 3\pi/16) + E(3\pi/16, -\pi/16)$$
$$- E(3\pi/16, 3\pi/16)|$$

Now, when measurement using B is done on the state $\frac{1}{\sqrt{2}}(|00\rangle - |11\rangle)$ above (or equivalently, measure $(R(\theta_1) \otimes R(\theta_2))\frac{1}{\sqrt{2}}(|00\rangle - |11\rangle)$ using the computational basis), recalling that $(R(\theta_1) \otimes R(\theta_2))\frac{1}{\sqrt{2}}(|00\rangle - |11\rangle) = \frac{1}{\sqrt{2}}(\cos(\theta_1 + \theta_2) |00\rangle + \sin(\theta_1 + \theta_2) |01\rangle + \sin(\theta_1 + \theta_2) |10\rangle - \cos(\theta_1 + \theta_2) |11\rangle)$, we have

$$P_{+1,+1}(\theta_1, \theta_2) = \frac{1}{2}\cos^2(\theta_1 + \theta_2)$$

---

4 Note that this basis B corresponds to the operator O

$$O = (R(\theta_1) \otimes R(\theta_2)) ((|0\rangle \langle 0| - |1\rangle \langle 1|) \otimes (|0\rangle \langle 0| - |1\rangle \langle 1|)) (R(\theta_1) \otimes R(\theta_2))^{-1}$$

with eigenstates which are the states in B. For example, we have $O(R(\theta_1) \otimes R(\theta_2) |00\rangle) = R(\theta_1) \otimes R(\theta_2) |00\rangle$, $O(R(\theta_1) \otimes R(\theta_2) |11\rangle) = R(\theta_1) \otimes R(\theta_2) |11\rangle$, $O(R(\theta_1) \otimes R(\theta_2) |01\rangle) = -R(\theta_1) \otimes R(\theta_2) |01\rangle$, and $O(R(\theta_1) \otimes R(\theta_2) |10\rangle) = -R(\theta_1) \otimes R(\theta_2) |10\rangle$.

$$P_{-1,-1}(\theta_1, \theta_2) = \frac{1}{2}cos^2(\theta_1 + \theta_2)$$

$$P_{+1,-1}(\theta_1, \theta_2) = \frac{1}{2}sin^2(\theta_1 + \theta_2)$$

$$P_{-1,+1}(\theta_1, \theta_2) = \frac{1}{2}sin^2(\theta_1 + \theta_2)$$

and so, $E(\theta_1, \theta_2) = cos^2(\theta_1 + \theta_2) - sin^2(\theta_1 + \theta_2)$, that is,

$$E(-\pi/16, -\pi/16) = cos^2(-\pi/8) - sin^2(-\pi/8) = \frac{1}{\sqrt{2}}$$

$$E(-\pi/16, 3\pi/16) = cos^2(\pi/8) - sin^2(\pi/8) = \frac{1}{\sqrt{2}}$$

$$E(3\pi/16, -\pi/16) = cos^2(\pi/8) - sin^2(\pi/8) = \frac{1}{\sqrt{2}}$$

$$E(3\pi/16, 3\pi/16) = cos^2(3\pi/8) - sin^2(3\pi/8) = -\frac{1}{\sqrt{2}}$$

Therefore, in the quantum case we have

$$S \approx \left| \frac{1}{\sqrt{2}} + \frac{1}{\sqrt{2}} + \frac{1}{\sqrt{2}} - (-\frac{1}{\sqrt{2}}) \right| = 2\sqrt{2}$$

that is, we can get $S > 2$ in the quantum case whereas the classical case has $S \leq 2$! The basis B is also the one that gives the maximum value for S, i.e. this is optimal in the quantum case [Buhrman et al., 2010], in a result called the Tsirelson's bound, i.e. $S(\theta_1, \theta_2) \leq 2\sqrt{2}$ in the quantum case, where

$$S(\theta_1, \theta_2) = |E(\theta_1, \theta_1) + E(\theta_1, \theta_2) + E(\theta_2, \theta_1) - E(\theta_2, \theta_2)|$$

with $S(\theta_1 = -\pi/16, \theta_2 = 3\pi/16) = 2\sqrt{2}$, as we showed above.
Recalling that

$$E(-\pi/16, -\pi/16)$$
$$= (+1)(+1) \cdot P_{+1,+1}(-\pi/16, -\pi/16) + (-1)(-1) \cdot P_{-1,-1}(-\pi/16, -\pi/16)$$
$$+ (+1)(-1) \cdot P_{+1,-1}(-\pi/16, -\pi/16) + (-1)(+1) \cdot P_{-1,+1}(-\pi/16, -\pi/16)$$

an experiment using the algorithm above running N_{00} rounds using settings $(-\pi/16, -\pi/16)$ could count the number of rounds where we get each output $(+1,+1)$, $(-1,-1)$, $(+1,-1)$, and $(-1,+1)$, and estimate for $(-\pi/16, -\pi/16)$ the value of E as follows, denoting by $N_{00}^{a,b}$ the number of occurrences of getting outputs $A_0 = a$ and $B_0 = b$ given inputs 00:

$$E(-\pi/16, -\pi/16) \approx \frac{N_{00}^{+1,+1} + N_{00}^{-1,-1} - N_{00}^{+1,-1} - N_{00}^{-1,+1}}{N_{00}}$$

$$= \frac{1}{N_{00}} \sum_{\nu=1}^{N_{00}} A_0^\nu B_0^\nu = \langle A_0 B_0 \rangle$$

where we used the definition of $\langle A_i B_j \rangle$ we saw earlier. We can get estimates for $E(-\pi/16, 3\pi/16)$, $E(3\pi/16, -\pi/16)$ and $E(3\pi/16, 3\pi/16)$ in a similar way and compute $S(\theta_1 = -\pi/16, \theta_2 = 3\pi/16)$ experimentally. Violations of the CHSH inequality have been experimentally shown as discussed in Wolf [2021].

3.1.2.3 Discussion

We can see from the above that there is something remarkable with entanglement that allows certain "no communication" problems to be solved that classically cannot be solved, by allowing certain correlated bits to be shared. In fact, the CHSH problem can be used as a test of genuine entanglement – if A and B can share genuine entangled pairs of qubits, then they should be able to solve the CHSH problem with higher probability than 0.75! In a way, entanglement seems to have made up (at least in part) for the "no communication" constraint between A and B.

3.2 Other Quantum Protocols

We briefly consider several other quantum versions of distributed system protocols, including quantum coin flipping, quantum leader election, quantum key distribution, quantum voting, quantum oblivious transfer, and the quantum Byzantine Generals solution.

3.2.1 Quantum Coin Flipping

3.2.1.1 Classical Coin Flipping

Coin flipping (or tossing) is the following problem first proposed by Blum [1983]. Before introducing quantum coin flipping, let us discuss the classical coin flipping problem first.[5]

Suppose machines A and B don't trust each other (or two people, typically Alice and Bob who is divorced and don't trust each other anymore) but wants to agree on a fair coin flip over the telephone. For example, machine A chooses "head" (or 1) and B chooses "tail" (or 0), and A flips a (presumably non-bias) coin and tells B over the phone that it is "head" – if B trusts A, then this is not a problem, but if not,

---

5 Parts of our discussion here uses the explanations from https://zoo.cs.yale.edu/classes/cs467/2013s/lectures/ln21.pdf [last accessed: 9/11/2022].

then how can B ensure that what A said is true? One approach is for both to first choose a side (e.g. A chooses 0 and B chooses 1), and then flip a coin each, and then both compute the winner as $a \oplus b$, where the probability of $a \oplus b = 0$ is 1/2 (since $a \oplus b = 0$ only when $a = b$ which is half the time) and the probability of $a \oplus b = 1$ is 1/2 (since $a \oplus b = 1$ only when $a \neq b$ which is half the time). Note that the goal of a coin flipping protocol is either both computes 0 or both computes 1, in either case, at the end, both should get the same result (the result of the coin flip), and that this coin flip should be "fair" in some sense – intuitively, neither A or B should be able to bias the result (in the sense of making itself or even the other party win!).

But a problem is that, *over the phone*, A and B need to tell each other their outcomes which might not happen concurrently, i.e. if A tells B its outcome first, then B, knowing a, can name a value for b in its favor, and similarly, if B tells A its outcome first; e.g. if A chose 0 and B chose 1, then suppose B receives the result from A first (say 0), then B can choose 1 and tell A its result is 1 (regardless of the actual outcome of its flip) and then both compute the winner as $a \oplus b = 0 \oplus 1 = 1$, so that B forces a win! Similarly, if A says 1, B tells A its flip result is 0 and forces a win, or if A receives the flip result from B first, A can force a win – i.e. it is not *secure*, or either player can force a result.

Of course, a trusted mediator can help here, who gets the result from both A and B before announcing the result $a \oplus b$. Note that this works because B provides its outcome to the mediator without knowing A's outcome. The key idea here is the notion of *bit commitment*, that is, suppose a is sent to the mediator first (without revealing anything to B), then A has committed to a value for a before knowing b, and then B, not knowing a, provides an outcome to the mediator, and cannot change its outcome after that – B also has committed to a value for b before knowing a. But suppose there is no such mediator – how would it work? We can use a cryptographic approach to perform such bit commitments, as in the idea in Blum's solution, which assumes parties are computationally bounded so that the protocol is secure (no party can cheat). Rather than presenting exactly Blum's solution,[6] we discuss a solution based on the more abstract *generic bit commitment protocol*. First, we explain the notion of bit commitment, without a trusted mediator.

Suppose we use encryption to do this commitment. One (naive) way is for machine A to flip a coin and the outcome a is encrypted using key k_A in a message and the encrypted message $E_{k_A}(a)$ is sent to B – this seems like a way for A to commit to an outcome without revealing it to B, but we shall see if this works. Machine B is unable to decrypt the message without A's key and flips its coin to get an outcome b which is then revealed to A. At this point, A can compute $a \oplus b$. Machine A then gives B the key k_A so B can decrypt to find a, and B can also now

6 A concise description is at https://xiaohuiliu.medium.com/another-fair-bitcoin-toss-742894b086cd [last accessed: 17/5/2022].

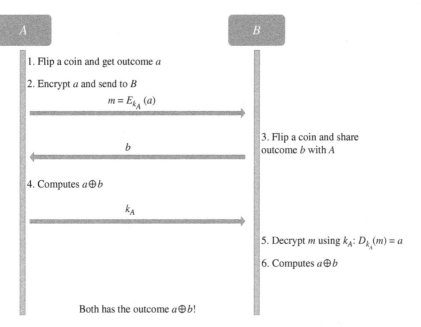

Figure 3.3 A solution to the coin flipping protocol without a trusted mediator, but has a problem. Note that before this protocol starts, *A* and *B* each has chosen their sides, and then the protocol starts to decide who wins.

compute $a \oplus b$. Both A and B now has the result. This protocol is illustrated in Figure 3.3.

But what is the problem? Note that A knows the outcome b before B gets A's key. Is there a way for A, now knowing B's outcome, to manipulate its encrypted outcome $E_{k_a}(a)$? It seems the answer is "yes"; if A makes use of a *colliding triple*, where a colliding triple (m, k_0, k_1) has the property that $D_{k_0}(m) = 0$ and $D_{k_1}(m) = 1$, that is, the same message m can be decrypted in two different ways, with two different keys, to get either 0 or 1. So, for example, A, knowing b, can force a favorable outcome by giving B either k_0 or k_1 so that B decrypts $E_{k_a}(a)$ to 0 or 1!

An example of such a colliding triple is $(m, m \oplus 0, m \oplus 1)$, with the decryption function: $D_k(m) = m \oplus k$. Using such a colliding triple, A sends the encrypted message $m = E_{k_a}(a)$ to B, and then if A wants m to decrypt to 0, A sends the key $m \oplus 0$, so that B decrypts m as follows: $D_{m \oplus 0}(m) = m \oplus (m \oplus 0) = 0$, and if A wants m to decrypt to 1, then A sends the key $m \oplus 1$, so that B decrypts m as follows: $D_{m \oplus 1}(m) = m \oplus (m \oplus 1) = 1$! So, A can choose whatever outcome it wants by sending the corresponding key to B! How can this be resolved?

One way to resolve this issue is for B to also play a role in choosing the key since A would not now be able to choose keys from a colliding triple. So, now instead of

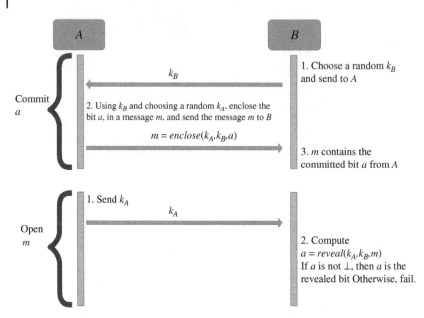

Figure 3.4 A generic bit commitment protocol.

only A supplying the key, machine A supplies k_A and B supplies k_B, and the key pair (k_A, k_B) is used to enclose (or encrypt) A's outcome and also for B to open (or decrypt) the encrypted outcome from A. Further, for this to work, we have these two properties:

1. so that B cannot see A's outcome (say in a message m) before sharing its outcome to A, using k_B alone cannot (or is hard to) open m, and
2. so that A cannot choose what outcome for m to decrypt to, for whatever k_A from A, B, using (k_A, k_B) will decrypt m to the same outcome.

We now have this bit commitment protocol as shown in Figure 3.4.

There are different possible schemes for realizing such a protocol; we sketch one method here using a hash function H for A to commit a. B chooses a random n-bit string k_B and sends this to A, which then chooses a random n-bit string k_A and computes $m = enclose(k_A, k_B, a) = H(k_B k_A a)$, i.e. applying the hash function H on the concatenation of k_B, k_A, and a. Machine A then sends m to B as its commitment.

Then, to open m, A sends a key k to B, which uses some algorithm[7] to find a' such that $m = H(k_B k a')$, where if a' is found (when k is the right key $k = k_A$), then $a' = a$ (given the collision-free property of the hash function that if $a' \neq a$, i.e. $k_B k_A a' \neq$

---

7 Note that, given k_B and k, to find a', B can compute the values $H(k_B k 0)$ and $H(k_B k 1)$ and see which value matches m.

$k_B k_A a$, then $H(k_B k_A a') \neq H(k_B k_A a))$, and we have the revealed bit; otherwise, if no such a' is found, then the protocol fails. Note that such an algorithm is the *reveal*(k_A, k_B, a) function.

Note that this protocol satisfies the properties we want above. For example, for (1) above, without k_A, given m, if B wants to find a, then B has to find k'_A and a' such that $m = H(k_B k'_A a')$, which would be like inverting H, and so, computationally hard given that H is typically a one-way function and for large enough n. This prevents B from extracting a from m before it shares its bit with A, and so, B cannot bias or force the result.

Also, for (2) above, in order for A to determine what m will decrypt to, depending on its key, A will need to come up with k_A and k'_A such that $H(k_B k_A 0) = H(k_B k'_A 1) = m$, so that suppose A wants m to decrypt to 0, it will send the key $k = k_A$ to B which uses its *reveal* algorithm to compute 0 (which satisfies $H(k_B k 0) = m$), and similarly, when A wants m to decrypt to 1, it will send B the key k'_A, which satisfies $H(k_B k'_A 1) = m$, so that B computes 1. But given the collision-free property of the hash function H, it is hard for A to come up with k_A and k'_A such that $H(k_B k_A 0) = H(k_B k'_A 1)$. This prevents A from determining its a after knowing B's value b, and so, A cannot bias/force the result.

With the bit commitment protocol that works, we can then allow A to commit to a bit without being able to dictate what it is after knowing B's value. The coin flipping protocol using bit commitment is given in Figure 3.5, with the output being $a \oplus b$ for both. Note that the output of the coin flipping protocol is either both have 0 or both have 1. And this protocol looks secure as long as the parties are computationally bounded, e.g. unable to invert the Hash function.

What further problems would there be with this? Note that after step 6. when A has computed $a \oplus b$, it might realize that it has lost and then what A could do is to stop the protocol and refuse to send the key to B! With the protocol terminating prematurely, it seems no one is the winner, or at least, B has been denied a win!

In general, one can state the properties of a coin flipping protocol and quantify the security of a protocol using a parameter value ϵ in the following sense. A coin flipping protocol is *correct* if when both parties (A and B participating in the protocol) are honest (follows the protocol), then the probability of the output o of the protocol being 1 is 1/2 and the probability of the output o being 0 is 1/2, i.e. there is equal chance of either party winning, or $prob(o = 0) = prob(o = 1) = 1/2$. A coin flipping protocol (the *strong* version) is said to be *ϵ-secure* when neither party can force the probability of the final outcome beyond certain bounds defined by ϵ (called the *bias* of the protocol). More precisely,

1. whatever A does (assuming B follows the protocol honestly), $prob(o = 0) \leq 1/2 + \epsilon$ and $prob(o = 1) \leq 1/2 + \epsilon$, i.e. A (in trying to make the coin flip bias toward 0) can do nothing (within some predefined space of strategies) to make

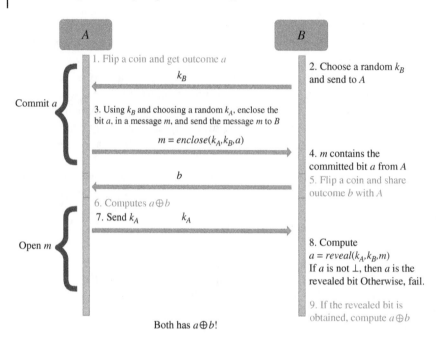

Figure 3.5 Coin flipping protocol using the bit commitment protocol – note the additions to the generic bit commitment protocol to realize coin flipping.

$prob(o = 0) > 1/2 + \epsilon$ and A (even in trying to make B win, for some reason) can do nothing to make $prob(o = 1) > 1/2 + \epsilon$;

2. (and similarly for B) whatever B does (assuming A follows the protocol honestly), $prob(o = 0) \leq 1/2 + \epsilon$ and $prob(o = 1) \leq 1/2 + \epsilon$, i.e. B (in trying to make the coin flip result bias toward 1) can do nothing (within some predefined space of strategies) to make $prob(o = 1) > 1/2 + \epsilon$ and B (even in trying to make A win, for some reason) can do nothing to make $prob(o = 0) > 1/2 + \epsilon$.

For the coin flipping protocol (the *weak* version), we just have the requirement that neither player can force the output to be more favorable toward itself, that is, assuming A wins if the outcome is 0, and B wins if the outcome is 1,

1. whatever A does, $prob(o = 0) \leq 1/2 + \epsilon$, i.e. A (in trying to make the coin flip bias toward 0) can do nothing to make $prob(o = 0) > 1/2 + \epsilon$,
2. (and similarly for B) whatever B does, $prob(o = 1) \leq 1/2 + \epsilon$, i.e. B (in trying to make the coin flip result bias toward 1) can do nothing to make $prob(o = 1) > 1/2 + \epsilon$.

Note that from the above two conditions $prob(o = 0) \leq 1/2 + \epsilon$ and $prob(o = 1)$ $\leq 1/2 + \epsilon$, we can write: $\epsilon = max(P_A^*, P_B^*) - \frac{1}{2}$, where P_A^* and P_B^* are the maximum winning probabilities of A and B, respectively.

Also, note that if one has a protocol where ϵ is small, say $\epsilon = 0.01$, then it seems "more secure," since, for example, A cannot make the $prob(o = 0) > 0.51$, whereas for a protocol with larger ϵ, say $\epsilon = 0.21$, A cannot make the $prob(o = 0) > 0.71$, but might be able to make the $prob(o = 0) = 0.7 \leq (0.5 + 0.21)$, which is still quite bias! Note that an insecure protocol (like the naive first one we saw), the party that receives the flip result from the other party first, can force a win, that is, if A gets B's result first, it can make $prob(o = 0) = 1.0 \leq (0.5 + 0.5)$! That is, it is $\epsilon = 1/2$-secure. If there would be an $\epsilon = 0$-secure protocol, then it is not only correct but "totally" secure. So, the smaller the ϵ the better is the protocol.

Then, how low can ϵ get? Note that the range for ϵ is $0 \leq \epsilon \leq 1/2$ since we are dealing with probabilities, but are there classical coin flipping protocols with very much lower ϵ, say some $\epsilon < 1/2$? It turns out that no strong classical coin flipping protocol can have $\epsilon < 1/2$ [Döscher and Keyl, 2002][8]! This seems contradictory to what we saw earlier with coin flipping using bit commitment achieved via the Hash function, where we achieved a "pretty good" protocol that seems "rather secure" – but note that the security of the protocol relies on parties being computationally bounded to be secure, or on not being able to quickly solve computationally hard problems such as inverting a hash function (this has been called *computational security*)! In a world where the parties have adequate (or unlimited) computational ability or plenty of time, then such a protocol is not secure (i.e. we say it is not *information-theoretically secure*), and in such a world, more precisely, no classical strong coin flipping protocol can have $\epsilon < 1/2$, i.e. $\epsilon = 1/2$, which means that a party can (even if the other cannot) indeed completely bias the outcome of the protocol. But for weak coin flipping, one can obtain arbitrarily small (non-zero) ϵ [Mochon, 2007; Aharonov et al., 2016].

What if we use quantum information – would that help? It turns out that a strong coin flipping protocol using quantum information can achieve a value for $\epsilon < 1/2$, but the smallest ϵ possible with any quantum strong coin flipping protocol is $\epsilon = (\sqrt{2} - 1)/2 \approx 0.207$ as shown by Kitaev, since such a protocol must satisfy $P_A^* \cdot P_B^* \geq 1/2$.[9] This means that even with quantum information, we cannot get a strong coin flipping protocol with $\epsilon = 0$!

There are other ways to solve the classical coin-flipping problem in practical ways which can be secure via computational bounds, and many variants of the problem, but we will not discuss these further here, and instead, go ahead to look

---

8 https://www.quantiki.org/wiki/strong-coin-tossing.
9 See the excellent Week8 notes by Wehner and Vidick at https://ocw.tudelft.nl/courses/quantum-cryptography.

at a quantum coin flipping protocol (a coin flipping protocol that uses quantum information), with an $0 < \epsilon < 1/2$.

3.2.1.2 Quantum Coin Flipping

Here, we use the quantum coin flipping protocol from Döscher and Keyl [2002], which uses qubits and qutrits (a three level quantum system, which can be in a superposition of three orthogonal states, $|0\rangle$, $|1\rangle$ and $|2\rangle$).

1. Party A flips a coin with the result $a \in \{0, 1\}$, which is stored, and according to this result, a quantum state is prepared in a quantum system involving a qubit q_s and a qutrit q_t:
   - if $a = 0$, then the state prepared is $|\psi_0\rangle = \frac{1}{\sqrt{2}}(|0, 0\rangle + |1, 2\rangle)$, and
   - if $a = 1$, then the state prepared is $|\psi_1\rangle = \frac{1}{\sqrt{2}}(|1, 1\rangle + |0, 2\rangle)$.

   (Note that the states $|\psi_0\rangle$ and $|\psi_1\rangle$ are orthogonal to each other, and so, can encode the two possible values of a.) Party A then sends parts of its quantum system's state, i.e. the qutrit q_t, to B, retaining the qubit state.
2. B receives the qutrit state and stores it. Also, B flips a coin with result $b \in \{0, 1\}$ and sends the value of b to A.
3. Party A receives b and sends the qubit q_s state to B. Party A can then already compute $a \oplus b$ at this point.
4. B now has the full state of A's quantum system $q_s q_t$ and performs the projective measurement with operators $P_0 = |\psi_0\rangle \langle\psi_0|$, $P_1 = |\psi_1\rangle \langle\psi_1|$ and $P_\emptyset = \mathbf{I} - P_0 - P_1$, where $\mathbf{I}$ is the identity matrix. If the protocol was followed as above, given the way A's quantum system was prepared, the full state of A's quantum system (as obtained by B) will be either $|\psi_0\rangle$ (corresponding to $a = 0$) or $|\psi_1\rangle$ (corresponding to $a = 1$), and the measurement will reveal the state $|\psi_0\rangle$ with high probability if a had been 0, or the state $|\psi_1\rangle$ with high probability if a had been 1 due to (1) above, obtaining a result $a' = a$. Party B can then already compute $a' \oplus b$, which should be equal to $a \oplus b$ obtained by A.

In the above, if both were honest, both can compute $a \oplus b$, which is either 0 or 1 with equal probability – the protocol is correct.

Is the protocol secure? Note that A's sending of its qutrit state to B in step (1) is in a way doing something like a bit commitment (though not quite as cheating is possible to an extent), and in step (3), when B receives q_s is like a reveal, since it now has both q_s and q_t. Party A can try to cheat, however, by sending B some qutrit and some qubit that works out in its favor. For example, suppose A chose $a = 0$, and now tries to influence the outcome to be 0 (in A's favor), as follows. Instead of $|\psi_0\rangle$, A uses the state

$$|\psi_0'\rangle = \frac{1}{\sqrt{6}}(|0, 0\rangle + |0, 1\rangle + 2\,|1, 2\rangle)$$

without B knowing about this change. Then, A continues till step (3) to receive b from B. And b can either be 0 or 1:

- Now, if the received $b = 0$, then the protocol continues without change, and the probability that B measures a' to be 0 is

$$tr\left(|\psi_0\rangle\langle\psi_0|\,|\psi_0'\rangle\langle\psi_0'|\right) = tr\left(\langle\psi_0|\psi_0'\rangle\langle\psi_0'|\psi_0\rangle\right) = \left|\langle\psi_0|\psi_0'\rangle\right|^2$$

$$= \left(\frac{1}{\sqrt{12}} + \frac{2}{\sqrt{12}}\right)^2 = \frac{3}{4}$$

That is, the probability of the outcome $a' \oplus b = 0 \oplus 0 = 0$ is 3/4.
- And if the received $b = 1$, then A applies an operator to the qubit q_s which swaps $|0\rangle$ and $|1\rangle$, before sending it to B. This means that B will have the state $q_s q_t$ as follows:

$$|\psi_1'\rangle = \frac{1}{\sqrt{6}}(|1,0\rangle + |1,1\rangle + 2\,|0,2\rangle)$$

And the probability that B measures a' to be 1 is

$$tr\left(|\psi_1\rangle\langle\psi_1|\,|\psi_1'\rangle\langle\psi_1'|\right) = tr\left(\langle\psi_1|\psi_1'\rangle\langle\psi_1'|\psi_1\rangle\right) = \left|\langle\psi_1|\psi_1'\rangle\right|^2$$

$$= \left(\frac{1}{\sqrt{12}} + \frac{2}{\sqrt{12}}\right)^2 = \frac{3}{4}$$

That is, the probability of the outcome $a' \oplus b = 1 \oplus 1 = 0$ is 3/4.

From the above, A has managed to bias the outcome beyond the uniform probability whatever the value of b turns out to be (assuming B is following the protocol honestly), so that the total probability of the outcome being 0 is $1/2 \cdot 3/4 + 1/2 \cdot 3/4 = 3/4$!

What about B? Can B try to cheat, assuming that B is trying to make the outcome 1? On receiving the state of q_t from A in step (2), B might try to work out what value of a is encoded in it by performing the measurement $\{|0\rangle\langle 0|, |1\rangle\langle 1|, |2\rangle\langle 2|\}$. Consider the following two cases:

- *Case 1*: If the measurement result (call this a'') is 0 or 1, i.e. not 2, then B knows what value of a is: if $a'' = 0$, then $a = 0$ (looking at the term $|0,0\rangle$ in $|\psi_0\rangle$), and if $a'' = 1$, then $a = 1$ (looking at the term $|1,1\rangle$ in $|\psi_1\rangle$), that is, $a'' = a$. Then, to get the outcome to be 1, B can set its $b = a'' \oplus 1$ and send this value to A, which computes $a \oplus b = a \oplus (a'' \oplus 1) = 1$. What is the probability that the measurement result $a'' \neq 2$? Looking at the amplitudes in $|\psi_0\rangle$ and $|\psi_1\rangle$, this happens with probability $(\frac{1}{\sqrt{2}})^2 = 1/2$ when the state sent was $|\psi_0\rangle$, and with probability $(\frac{1}{\sqrt{2}})^2 = 1/2$ when the state sent was $|\psi_1\rangle$. This means that whether A sends $|\psi_0\rangle$ or $|\psi_1\rangle$, if $a'' \neq 2$, B can bias the result to get 1 for sure, and $a'' \neq 2$ happens with probability 1/2.

- *Case 2*: If the measurement result $a'' = 2$, then B is not able to learn anything about the value of a, and so continues with the protocol as normal and gets the desired result 1 with probability 1/2.

From the above, B has managed to bias the outcome beyond the uniform probability, i.e. since case 1 occurs with probability 1/2, where B is sure to get the outcome 1, and case 2 occurs with probability 1/2, where B gets 1 with probability 1/2, the total probability of the outcome being 1 is $1/2 \cdot 1 + 1/2 \cdot 1/2 = 3/4$!

And this is indeed the minimum bias that can be achieved in this protocol [Döscher and Keyl, 2002], that is, we have $\epsilon = 0.25 < 1/2$ for this protocol. Is there a protocol with a lower bias? As noted earlier, Kitaev showed that the best possible bias a strong coin flipping protocol can have is $(\sqrt{2} - 1)/2 \approx 0.207$, that is, there could be a strong coin flipping protocol with lower bias. Indeed, Chailloux and Kerenidis [2009] showed how a strong coin flipping protocol with bias $(\sqrt{2} - 1)/2 + O(\epsilon)$ can be constructed, using any weak coin-flipping protocol with bias ϵ. But for weak coin flipping, we noted earlier that one can obtain arbitrarily small (nonzero) ϵ as Mochon showed, and so, Chailloux and Kerenidis's method can provide a strong coin flipping protocol with bias arbitrarily close to the optimal $(\sqrt{2} - 1)/2$.

We do not provide a comprehensive survey of quantum coin flipping here, but the interested reader can look into the body of work on quantum weak coin flipping, including weak coin flipping protocols by Kerenidis and Nayak [2004], Ambainis [2004], and Spekkens and Rudolph [2002] with bias less than 0.25, one by Mochon with bias 0.192 [Mochon, 2004], a family of protocols with bias 1/6 also by Mochon [2005], and a protocol with bias 1/10 [Arora et al., 2019], using some nontrivial mathematical techniques.

There are variations of quantum coin flipping protocols such as coin flipping for more than two parties [Ambainis et al., 2004], which uses a weak two-party coin flipping protocol between pairs of parties in multiple rounds, and a strong coin flipping protocol for the final round to decide the final outcome.

Quantum coin flipping protocols have been implemented experimentally as far back as 2011 and 2014 (e.g. see [Berlín et al., 2011; Pappa et al., 2014]).

Note that the difficulty is due to the lack of a trusted mediator, and that A and B don't trust each other. If there was a trusted mediator, it could send to each of A and B one qubit of a shared entangle pair of qubits (say, $\frac{1}{\sqrt{2}}(|00\rangle + |11\rangle)$), and then when they measure their qubits, there is equal probability of both getting 0 or both getting 1. For multiparty (more than two parties) coin flipping, with a trusted mediator, one could use a generalized GHZ state, $\frac{1}{\sqrt{2}}(|0 \ldots 0\rangle + |1 \ldots 1\rangle)$.

3.2.2 Quantum Leader Election

We consider the leader election problem, which is to elect a leader among a collection of distributed parties. Typically, in practical settings, one way to solve this is for each party to use its unique identifier, e.g. an IP address, and there is a protocol to solve this problem by selecting the party with the largest identifier as the leader (based on some ordering). However, when no such unique identifier can be established, then it is more challenging. One way to solve this problem is for each party to run some subroutine for a number of rounds, with each round filtering out some parties to be ineligible, starting with all parties being eligible. For example, in each round, each eligible party generates a random bit (0 or 1), and those with bit 0 are filtered out as ineligible. Eventually, there could remain only one eligible party who becomes the leader. However, such a protocol is probabilistic – suppose, in a round, every eligible party gets a 1, and no one is filtered out, then one can keeping repeating this, each party generates a new bit and see if any parties can be filtered out, and the process might be repeated a number of times, and still no one is eliminated – we depend on chance that eventually someone gets eliminated, and so this method is not exact (in the sense that one cannot exactly say when the leader will eventually be elected, e.g. with bounded time – there could be many rounds and repetitions if one is unlucky so that no one gets eliminated for a while). In fact, this problem without unique identifiers (so-called *anonymous networks*) has no classical exact solution for many network topologies in which the parties are connected [Tani et al., 2012]. Also, since there are no cheaters assumed, this is called the *fair* leader election problem. However, can this problem be solved exactly with quantum information? The answer is yes, and Tani et al. [2012] provide an exact quantum algorithm for this fair leader election problem that elects a leader in polynomial time/communication complexity, showing another difference between the classical and quantum worlds.

In the following, we consider the leader election problem with cheaters and roughly sketch how one can use the coin flipping protocol to solve this problem – the algorithm is from Ganz [2017].

For simplicity, we just sketch the idea for $n = 2^k$ parties, that is, n is some power of two. Ganz's algorithm is more general [Ganz, 2017].

With potential cheaters, the leader election problem is to elect a leader where:

a. if all parties are honest, then each party has a $\frac{1}{n}$ chance of being the leader, i.e. anyone has an equal chance of being the leader,

b. otherwise (if there are cheaters), any honest player has probability $> \frac{1}{n} - \epsilon$ of being the leader.

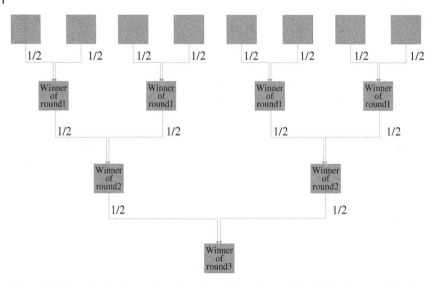

Figure 3.6 Leader election where the winner of each pair is decided using a weak coin flipping protocol. Honest parties shown, where the winning probability is 1/2 in each round. The leader has to be "lucky enough" to win in all three rounds and does so with probability $(\frac{1}{2})^3$.

Clearly, a protocol with small ϵ is better. The idea to solving the above problem is to carry out a knock-out tournament using the weak coin flipping protocol between pairs of parties and the winner of the tournament is declared the leader.

For (a), when all the parties are honest, then $k = log_2 \, n$ rounds of the tournament will determine the leader. To be leader, a party has to win all the k rounds, which it does so with probability $(\frac{1}{2})^k = \frac{1}{n}$, as illustrated in Figure 3.6 for $n = 2^3$.

For (b), using a weak (quantum) coin flipping protocol, with cheaters, in each round, in a pair playing this protocol, the cheater could win with probability at most $\frac{1}{2} + \epsilon$, i.e. an honest player can win with probability at least $1 - (\frac{1}{2} + \epsilon) = \frac{1}{2} - \epsilon$. Hence, in total, after $k = log_2 \, n$ rounds, the probability that the honest player (honest in all rounds) can become the leader is at least:

$$\left(\frac{1}{2} - \epsilon\right)^k = \left(\frac{1 - 2\epsilon}{2}\right)^k \geq \frac{1 - 2\epsilon k}{2^k} = \frac{1}{n} - \epsilon\frac{2log_2 \, n}{n} \geq \frac{1}{n} - \epsilon$$

using $(1 + x)^n \geq 1 + nx$, for $x \geq -1$, $n \in \mathbb{N}$, which applies since $x = -2\epsilon$ and $\epsilon \leq 1/2$. Note that the running time can be characterized by $O(log \, n)$ rounds, and with each round taking time R in a weak coin flipping protocol with bias ϵ, we have a total running time of $O(R \cdot log \, n)$.

But if n is not a power of two, further steps are required. A recursive algorithm called *Leader*($\{A_1, \dots, A_n\}, \epsilon$) to elect a leader from among n parties $\{A_1, \dots, A_n\}$ with an honest player being the leader with probability at least $\frac{1}{n} - \epsilon$ is as follows.

Before giving the algorithm, we introduce some notation. So far, we have looked at only *balanced* weak coin flipping protocols, but there are *unbalanced* weak coin flipping protocols (with bias ϵ) where if both parties are honest, then one party will have probability q of winning and the other party will have probability $1 - q$ of winning (in contrast to the balanced protocols, where both parties have equal probability $1/2$ of winning). An unbalanced weak coin flipping protocol with bias ϵ is denoted as $W_{q,\epsilon}$. A balanced weak coin flipping protocol with bias ϵ is denoted as W_ϵ.

Leader$(\{A_1, \ldots, A_n\}, \epsilon)$:

Step 1: Find $k \in \mathbb{N}$ such that $2^k \leq n < 2^{k+1}$.

Step 2: The first 2^k players $\{A_1, \ldots, A_{2^k}\}$ play a tournament among themselves (k rounds) using $W_{\frac{\epsilon}{2}}$ to find a winner, denoted as l_1.

Step 3: Compute $l_2 = Leader(\{A_{2^k+1}, \ldots, A_n\}, \epsilon)$

Step 4: l_1 plays l_2 using a protocol $W_{\left(\frac{2^k}{n}\right)', \frac{\epsilon}{2}}$, where $W_{q', \frac{\epsilon}{2}}$ denotes a protocol such that $\left| q' - \frac{2^k}{n} \right| \leq \frac{\epsilon}{2}$. The winner is the leader.

We will not repeat here the deep analysis given by Ganz [2017], but comment that the protocol is essentially a divide-and-conquer approach with recursive calls, and an honest player A_i has probability of at least $\frac{1}{n} - \epsilon$ of being the leader.

For leader election, the challenge is not having a trusted mediator and possible cheaters. If there was a trusted mediator, the mediator can prepare the W state: $|W\rangle = \frac{1}{\sqrt{3}}(|001\rangle + |010\rangle + |100\rangle)$, or its generalizations: $|W'\rangle = \frac{1}{\sqrt{n}}(|000 \ldots 001\rangle + |000 \ldots 010\rangle + \cdots + |100 \ldots 000\rangle)$. And send a qubit to each party, and with n parties sharing the state, after measurement, the leader is the party with the bit 1.

3.2.3 Quantum Key Distribution (QKD)

Keys are bit strings used for securing data, using encryption and decryption. Assuming that two parties want to exchange messages so that one who intercepts these messages cannot understand what the messages are. The general idea is that a k-bit string, for some large enough k, is used in an algorithm (denoted here by *enc*) to encrypt a message, i.e. convert a message m into a form that looks like a random string (and unreadable), and then this encoded string (i.e. $enc(m, k)$) is sent to the destination which can then use an algorithm (denoted here by *dec*) to decrypt, using the same key k, the transmitted string to extract m, i.e. $dec(enc(m, k), k) = m$. For all this to work, both parties need to share the secret key k.

How can such a key k be generated and shared among the two parties? There are many (classical) ways to distribute such secret keys, among communicating parties. But here, we look at quantum approaches to this problem, which has been well-studied. As far back as 1984, Charles Bennett and Gilles Brassard introduced the BB84 protocol which uses quantum information to generate such a shared key among two communicating parties.[10] QKD is well described in other sources, e.g. see Wolf [2021] and Van Assche [2006], but we outline the BB84 protocol here, based mainly on the excellent description by Vidick and Wehner.[11] In this protocol, A basically generates the key and "communicates" it to B but in a way that is secure and an eavesdropper would not be able to get the key without being detected. We assume that there is a classical channel between A and B.

1. A chooses an n-bit string $x = x_1 \ldots x_n \in \{0,1\}^n$ uniformly at random. A also chooses a basis string $\theta = \theta_1 \ldots \theta_n$ uniformly at random, which determines what basis to encode each bit of x. A transmits x to B by encoding each bit in a quantum state as follows: $H^{\theta_j}|x_j\rangle$, where $\theta_j \in \{0,1\}$, i.e. A effectively sends n qubits to B. This means that for some x_j, if $\theta_j = 0$, the computational basis is used, and x_j is encoded as $|x_j\rangle$ and if $\theta_j = 1$, the Hadamard basis is used, and x_j is encoded as $H|x_j\rangle$ (recall that $H|0\rangle = |+\rangle = (|0\rangle + |1\rangle)/\sqrt{2}$ and $H|1\rangle = |-\rangle = (|0\rangle - |1\rangle)/\sqrt{2}$).
2. B chooses a basis string $\theta' = \theta'_1 \ldots \theta'_n$ uniformly at random and measures qubit j it received from A in the basis θ'_j to produce outcome x'_j.
3. B tells A over the classical channel that it has received and measured all the qubits.
4. A and B then tell each (over the classical channel) their basis strings θ and θ'.
5. A and B then discards all rounds where they did not use the same basis. The set of indices for rounds which are kept $S = \{j \mid \theta_j = \theta'_j\}$, and since A and B choose from two basis options with equal probability, we have for large n, roughly $n/2$ bits kept, discarding roughly half of the rounds.
6. To test for an eavesdropper, which might have intercepted qubits from A, measuring them and resending them to B, i.e. an intercept-resend attack, A picks a subset $T \subseteq S$ for testing, where $|T| = n/4$ bits, that is, roughly half the kept bits (which were measured in the same basis). Note that for the bits indexed in T, since they were the result of using the same basis (for encoding by A and measurement by B), we should expect all of these bits to be the same; if a significantly more of these bits are not the same (i.e. the qubits have

---

10 It is online at https://arxiv.org/pdf/2003.06557.pdf [last accessed: 9/10/2022].
11 See Week 6 of the notes at https://ocw.tudelft.nl/courses/quantum-cryptography/? view=readings in the course: https://online-learning.tudelft.nl/courses/quantum-cryptography [last accessed: 14/6/2022]. The reader is also referred to the newly released book by Vidick and Wehner entitled "Introduction to Quantum Cryptography" (2024), Cambridge University Press.

been "disturbed"), this suggests there has been an eavesdropper who did an intercept-resend attack – to see this, see **Aside**.

Aside: Consider an eavesdropper that *always measures in the computational basis* (there could be more sophisticated attackers but this is just an example), e.g. when A sends $|x_j\rangle$ (i.e. encoding the bit x_j in the computational basis), the eavesdropper measures in the same basis as A and suppose B also measures in the same basis, then the eavesdropper has got it right and will be undetected in this instance! But suppose $x_j = 0$, and A sends $H|x_j\rangle = |+\rangle = (|0\rangle + |1\rangle)/\sqrt{2}$ (i.e. encoding the bit 0 in the Hadamard basis), the eavesdropper measures in the computational basis and gets eigenvalue 1 say with $|0\rangle$ with probability 1/2 and −1 with $|1\rangle$ with probability 1/2. Suppose the eavesdropper gets $|0\rangle$ and forward this to B, and suppose B happens to measure in the same basis that A used, then, since $|0\rangle = (|+\rangle + |-\rangle)/\sqrt{2}$, B gets $|+\rangle$ with probability 1/2 and $|-\rangle$ with probability 1/2; now, suppose B gets $|-\rangle$, then the summary is that A and B used the same basis, but the result that B obtained is $|-\rangle$ which is different from what A sent! So the eavesdropping has affected the qubits so that the result is not what is expected – in this case, leading to errors that should not have happened. We can compute the statistics to see how much error such an eavesdropper E will introduce in all cases where A and B used the same basis (that is, when we do not expect an error):

A's bit	A's encoding	E's measured and resend state	B's result using the same basis as A	Error?
0	$|0\rangle$	$|0\rangle$	$|0\rangle$	N
1	$|1\rangle$	$|1\rangle$	$|1\rangle$	N
0	$|+\rangle$	$|0\rangle$	$|+\rangle$	N
			$|-\rangle$	Y
		$|1\rangle$	$|+\rangle$	N
			$|-\rangle$	Y
1	$|-\rangle$	$|0\rangle$	$|+\rangle$	Y
			$|-\rangle$	N
		$|1\rangle$	$|+\rangle$	Y
			$|-\rangle$	N

> One can see a lot more errors due to E; without E there would not be any errors when both A and B are using the same basis (ignoring noisy channels).

The test for an eavesdropper is done as follows: A and B announce (over the classical channel) to each other values x_j and x'_j where $j \in T$ and compute the error rate $err = |W|/|T|$, where $W = \{j \in T | x_j \neq x'_j\}$ is the set of indices where there is a mismatch even when A and B used the same basis. So, if the error rate $err > 0$, then A and B abort the protocol, since there is likely an eavesdropper. But if $err = 0$, then there is likely no eavesdropping (though that is no guarantee that it is so since the eavesdropper might indeed be super-lucky to be undetected!) the set of bits corresponding to those in S but not used for testing, that is, indices in $S \setminus T$ will be a sequence of shared bits that will be used by A and B to form the key, i.e. on A's side, the key is x_j where $j \in S \setminus T$, and on B's side, the key is x'_j where $j \in S \setminus T$. Note that $|S \setminus T|$ is around $n/4$. In fact due to possible errors in transmission, it may be alright to not abort as long as $err < \epsilon$ for some bound ϵ on the error rate. Note that we used $n/4$ bits for testing to check if the remaining $n/4$ bits would be usable – and the tested bits are not the ones used – how does this work? As explained in Wolf [2021], and we won't repeat the argument here, the error rate in the sample $T \subseteq S$ is, with high probability, close to the error rate for the whole string (of bits in) S, when n is large, i.e. a high error rate in the sample suggests a high error rate in the whole string, so that testing with the sample already provides an adequate picture of the error rate to expect with the bits in $S \setminus T$.

7. A and B can go through a process called *privacy amplification* (an example given below) to further improve security, just in case (e.g. since small errors due to transmission is difficult to be differentiated from errors due to eavesdropping), which would result in the final key k of length less than $|S \setminus T| = n/4$.

The idea of privacy amplification is to further improve the secrecy of keys by extracting a (smaller) "real key" (of length less than $n/4$) to be used from the remaining bits indexed in $S \setminus T$. One way to do simple privacy amplification is to take pairs of bits from $S \setminus T$ and XOR them (or add modulo two, i.e. applying $\oplus$) to obtain a shorter (but more secure key).[12] There are other more sophisticated privacy amplification methods [Wolf, 2021; Van Assche, 2006].

---

12 This algorithm is explained as follows, where A is Alice, B is Bob and the eavesdropper is Eve:

> After error correction, Alice and Bob have identical copies of a key, but Eve may still have some information about it... Alice and Bob thus need to reduce Eve's information to an arbitrarily low value using some privacy amplification protocols. These classical protocols typically work as follows. Alice again randomly chooses pairs of bits and

For error correction, there is also a step called *information reconciliation* which after the error rate has been determined, via the testing, A and B proceed to correcting the errors, as elaborated on in Wolf [2021]. A protocol to ensure that both parties have the same bit string, i.e. to perform "error key reconciliation" (called Cascade) using parity values on blocks of bit strings is described in Mehic et al. [2022].

We next consider the QKD protocol by Artur Ekert called the E91 protocol [Ekert, 1991] that uses entangled states (Bell states) shared among the two communicating parties to generate a shared key between them. The protocol proceeds as follows to generate an n-bit key shared by the two parties A and B.

1. A third party is a source of n entangled pairs of qubits, each in the state: $\frac{1}{\sqrt{2}}(|00\rangle + |11\rangle)$. Each of the n two-qubit states is then shared by the two parties by the third party sending one qubit of each pair to each party. (We can also use other states such as the singlet state as originally done in Ekert [1991] to a similar effect.)
2. On receiving the qubits, for each qubit received, A randomly chooses one basis from among three possibilities, denoted by $\mathbf{a_1}$, $\mathbf{a_2}$, and $\mathbf{a_3}$:
   - $\mathbf{a_1}$: A basis based on observable Z, i.e. a measurement with outcome either $+1$ corresponding to Z's eigenvector $|0\rangle$ or -1 corresponding to the other eigenvector $|1\rangle$ of Z,
   - $\mathbf{a_2}$: A basis based on observable $(X + Z)/\sqrt{2} = H$, with outcome $+1$ corresponding to H's eigenvector $|\times\rangle = cos(\frac{\pi}{8})|0\rangle + sin(\frac{\pi}{8})|1\rangle$ or -1 corresponding to the other eigenvector $|\div\rangle = -sin(\frac{\pi}{8})|0\rangle + cos(\frac{\pi}{8})|1\rangle)$ of H,[13] and

---

computes their XOR value. But, in contrast to error correction, she does not announce this XOR value. She only announces which bits she chose (e.g. bits number 103 and 537). Alice and Bob then replace the two bits by their XOR value. In this way, they shorten their key while keeping it error free, but if Eve has only partial information on the two bits, her information on the XOR value is even less. Assume, for example, that Eve knows only the value of the first bit and nothing about the second one. Then she has no information at all about the XOR value. Also, if Eve knows the value of both bits with 60% probability, then the probability that she correctly guesses the XOR value is only $0.6^2 + 0.4^2 = 52\%$. This process would have to be repeated several times; more efficient algorithms use larger blocks ... [Gisin et al., 2002, Section C4 in Arxiv version, https://arxiv.org/pdf/quant-ph/0101098.pdf, accessed: 22/11/2022]

Roughly, the resulting key using this way of privacy amplification might be of length $(n/4)/2 = n/8$.
13 One can check that these are the eigenvectors of H via trigonometric identities (e.g. see https://sciencenotes.org/trig-identities-study-sheet) and noting that $cos(\frac{\pi}{4}) = \frac{1}{\sqrt{2}}$.

- a_3: A basis based on observable X, with outcome $+1$ corresponding to X's eigenvector $|+\rangle = \frac{1}{\sqrt{2}}(|0\rangle + |1\rangle)$ or -1 corresponding to the other eigenvector $|-\rangle = \frac{1}{\sqrt{2}}(|0\rangle - |1\rangle)$ of X.

and measures the qubit in that basis and records this. Similarly, for each qubit received, B randomly chooses one basis from three possibilities: b_1: observable $(X + Z)/\sqrt{2}$, b_2: observable X, and b_3: observable $(X - Z)/\sqrt{2}$ (with outcome $+1$ corresponding to eigenvector $|\times'\rangle = sin(\frac{\pi}{8})|0\rangle + cos(\frac{\pi}{8})|1\rangle$ or -1 corresponding to the other eigenvector $|\div'\rangle = -cos(\frac{\pi}{8})|0\rangle + sin(\frac{\pi}{8})|1\rangle$), and measures the qubit in that basis.

All the measurement results are recorded including the basis used for each qubit measurement. Then, both parties announce to the other (via a classical channel) what basis was used for measuring each qubit.

3. Each party then categorizes the measurement results obtained into two categories, one where both parties measured in the same basis and another where they measured in different bases. Each party announces publicly the measurement results where measurements were done in different bases (those in the second category). (The bases used and the measurement results for those cases where different bases were used are needed for the test in the next step.)

4. To test whether there has been eavesdropping by a third party, say E, which would "disturb" the transmitted qubits – e.g. E could intercept a qubit (and measure to see what it is) and transmit some qubit in place of the intercepted one, both parties can compute statistics on the measurements done with different bases and perform an analysis, as we show later. If the qubits appear to be undisturbed, then the outcomes from qubits measured in the same basis by both parties will be used as the shared secret key – both parties will have exactly the same outcomes (for qubits measured in the same basis) and so each such obtained bit will be a shared secret bit among the two parties. So, with n entangled pairs of qubits, the parties will share an n-bit key.

We consider further how the test for "disturbance" by an intermediary mentioned above works. The test works by relying on a mathematical relation based on the probabilities of correlated and anticorrelated outcomes as computed (or as expected) from quantum mechanics, and whether the statistics obtained from measurements reflects what we would expect. If it does reflect, then the test is passed; fail, otherwise.

Let us construct a table showing the expected (according to quantum mechanics) probabilities of different outcomes depending on the basis used by A and B. Let $P_{+1,+1}(\mathbf{a_i}, \mathbf{b_j})$ denote the probability of A obtaining outcome $+1$ and B obtaining outcome $+1$, given that A used basis $\mathbf{a_i}$ and B used basis $\mathbf{b_j}$, and similarly, with $P_{-1,-1}(\mathbf{a_i}, \mathbf{b_j})$, $P_{+1,-1}(\mathbf{a_i}, \mathbf{b_j})$ and $P_{-1,+1}(\mathbf{a_i}, \mathbf{b_j})$. Let the relation E which depends on the choice of A's basis $\mathbf{a_i}$ and B's basis $\mathbf{b_j}$, be given by

$$E(\mathbf{a_i}, \mathbf{b_j}) = P_{+1,+1}(\mathbf{a_i}, \mathbf{b_j}) + P_{-1,-1}(\mathbf{a_i}, \mathbf{b_j}) - P_{+1,-1}(\mathbf{a_i}, \mathbf{b_j}) - P_{-1,+1}(\mathbf{a_i}, \mathbf{b_j})$$

which represents the correlation of the outcomes obtained when A and B measured with bases $\mathbf{a_i}$ and $\mathbf{b_j}$. The higher the value of E, the greater the correlation.

A's basis	B's basis	$P_{+1,+1}$	$P_{-1,-1}$	$P_{+1,-1}$	$P_{-1,+1}$	E
$\mathbf{a_1}: Z$	$\mathbf{b_1}: (X+Z)/\sqrt{2}$	$\frac{1}{2}\cos^2(\frac{\pi}{8})$	$\frac{1}{2}\cos^2(\frac{\pi}{8})$	$\frac{1}{2}\sin^2(\frac{\pi}{8})$	$\frac{1}{2}\sin^2(\frac{\pi}{8})$	$\frac{1}{\sqrt{2}}$
$\mathbf{a_1}: Z$	$\mathbf{b_2}: X$	$\frac{1}{2}\cdot\frac{1}{2}$	$\frac{1}{2}\cdot\frac{1}{2}$	$\frac{1}{2}\cdot\frac{1}{2}$	$\frac{1}{2}\cdot\frac{1}{2}$	0
$\mathbf{a_1}: Z$	$\mathbf{b_3}: (X-Z)/\sqrt{2}$	$\frac{1}{2}\sin^2(\frac{\pi}{8})$	$\frac{1}{2}\sin^2(\frac{\pi}{8})$	$\frac{1}{2}\cos^2(\frac{\pi}{8})$	$\frac{1}{2}\cos^2(\frac{\pi}{8})$	$-\frac{1}{\sqrt{2}}$
$\mathbf{a_2}: (X+Z)/\sqrt{2}$	$\mathbf{b_1}: (X+Z)/\sqrt{2}$	$\frac{1}{2}$	$\frac{1}{2}$	0	0	1
$\mathbf{a_2}: (X+Z)/\sqrt{2}$	$\mathbf{b_2}: X$	$\frac{1}{2}\cos^2(\frac{\pi}{8})$	$\frac{1}{2}\cos^2(\frac{\pi}{8})$	$\frac{1}{2}\sin^2(\frac{\pi}{8})$	$\frac{1}{2}\sin^2(\frac{\pi}{8})$	$\frac{1}{\sqrt{2}}$
$\mathbf{a_2}: (X+Z)/\sqrt{2}$	$\mathbf{b_3}: (X-Z)/\sqrt{2}$	$\frac{1}{2}\cdot\frac{1}{2}$	$\frac{1}{2}\cdot\frac{1}{2}$	$\frac{1}{2}\cdot\frac{1}{2}$	$\frac{1}{2}\cdot\frac{1}{2}$	0
$\mathbf{a_3}: X$	$\mathbf{b_1}: (X+Z)/\sqrt{2}$	$\frac{1}{2}\cos^2(\frac{\pi}{8})$	$\frac{1}{2}\cos^2(\frac{\pi}{8})$	$\frac{1}{2}\sin^2(\frac{\pi}{8})$	$\frac{1}{2}\sin^2(\frac{\pi}{8})$	$\frac{1}{\sqrt{2}}$
$\mathbf{a_3}: X$	$\mathbf{b_2}: X$	$\frac{1}{2}$	$\frac{1}{2}$	0	0	1
$\mathbf{a_3}: X$	$\mathbf{b_3}: (X-Z)/\sqrt{2}$	$\frac{1}{2}\cos^2(\frac{\pi}{8})$	$\frac{1}{2}\cos^2(\frac{\pi}{8})$	$\frac{1}{2}\sin^2(\frac{\pi}{8})$	$\frac{1}{2}\sin^2(\frac{\pi}{8})$	$\frac{1}{\sqrt{2}}$

The above probabilities were computed using a number of expressions and identities – see **Aside**.

Aside: The following observations are useful for computing the probabilities above and can be verified by the reader via calculations.

- We note that: $\frac{1}{\sqrt{2}}(|00\rangle + |11\rangle) = \frac{1}{\sqrt{2}}(|++\rangle + |--\rangle) = \frac{1}{\sqrt{2}}(|\times\times\rangle + |\div\div\rangle)$
- We note the different ways of expressing one basis in terms of the other:

$$|0\rangle = \frac{1}{\sqrt{2}}(|+\rangle + |-\rangle) \qquad |1\rangle = \frac{1}{\sqrt{2}}(|+\rangle - |-\rangle)$$

$$|0\rangle = \cos\left(\frac{\pi}{8}\right)|\times\rangle - \sin\left(\frac{\pi}{8}\right)|\div\rangle \qquad |1\rangle = \sin\left(\frac{\pi}{8}\right)|\times\rangle + \cos\left(\frac{\pi}{8}\right)|\div\rangle$$

$$|0\rangle = \sin\left(\frac{\pi}{8}\right)|\times'\rangle - \cos\left(\frac{\pi}{8}\right)|\div'\rangle \qquad |1\rangle = \cos\left(\frac{\pi}{8}\right)|\times'\rangle + \sin\left(\frac{\pi}{8}\right)|\div'\rangle$$

$$|\times\rangle = \cos\left(\frac{\pi}{8}\right)|+\rangle + \sin\left(\frac{\pi}{8}\right)|-\rangle \qquad |\div\rangle = \sin\left(\frac{\pi}{8}\right)|+\rangle - \cos\left(\frac{\pi}{8}\right)|-\rangle$$

$$|\times\rangle = \frac{1}{\sqrt{2}}(|\times'\rangle - |\div'\rangle) \qquad |\div\rangle = \frac{1}{\sqrt{2}}(|\times'\rangle + |\div'\rangle)$$

- We also note that $cos(\frac{\pi}{4}) = sin(\frac{\pi}{4}) = \frac{1}{\sqrt{2}}$ and use the identities:

$$\frac{1}{\sqrt{2}} \left(cos\left(\frac{\pi}{8}\right) + sin\left(\frac{\pi}{8}\right) \right) = cos\left(\frac{\pi}{8}\right);$$

$$\frac{1}{\sqrt{2}} \left(cos\left(\frac{\pi}{8}\right) - sin\left(\frac{\pi}{8}\right) \right) = sin\left(\frac{\pi}{8}\right)$$

So, for example, to compute the entries $P_{+1,+1}(\mathbf{a_1}, \mathbf{b_1})$, $P_{-1,-1}(\mathbf{a_1}, \mathbf{b_1})$, $P_{+1,-1}(\mathbf{a_1}, \mathbf{b_1})$ and $P_{-1,+1}(\mathbf{a_1}, \mathbf{b_1})$, in the first row, we can write $\frac{1}{\sqrt{2}}(|00\rangle + |11\rangle)$ as

$$\frac{1}{\sqrt{2}} \left(|0\rangle \left(cos\left(\frac{\pi}{8}\right) |\times\rangle - sin\left(\frac{\pi}{8}\right) |\div\rangle \right) + |1\rangle \left(sin\left(\frac{\pi}{8}\right) |\times\rangle + cos\left(\frac{\pi}{8}\right) |\div\rangle \right) \right)$$

$$= \frac{1}{\sqrt{2}} cos\left(\frac{\pi}{8}\right) |0\rangle |\times\rangle - \frac{1}{\sqrt{2}} sin\left(\frac{\pi}{8}\right) |0\rangle |\div\rangle + \frac{1}{\sqrt{2}} sin\left(\frac{\pi}{8}\right) |1\rangle |\times\rangle$$

$$+ \frac{1}{\sqrt{2}} cos\left(\frac{\pi}{8}\right) |1\rangle |\div\rangle$$

Then, to get $(+1,+1)$, we need to obtain $|0\rangle |\times\rangle$ which occurs with probability $(\frac{1}{\sqrt{2}} cos(\frac{\pi}{8}))^2$, to get $(-1,-1)$, we need to obtain $|1\rangle |\div\rangle$ which occurs with probability $(\frac{1}{\sqrt{2}} cos(\frac{\pi}{8}))^2$, to get $(+1,-1)$, we need to obtain $|0\rangle |\div\rangle$ which occurs with probability $(-\frac{1}{\sqrt{2}} sin(\frac{\pi}{8}))^2$, and to get $(-1,+1)$, we need to obtain $|1\rangle |\times\rangle$ which occurs with probability $(\frac{1}{\sqrt{2}} sin(\frac{\pi}{8}))^2$.

We can then compute the expression[14]:

$$S = E(\mathbf{a_1}, \mathbf{b_1}) - E(\mathbf{a_1}, \mathbf{b_3}) + E(\mathbf{a_3}, \mathbf{b_1}) + E(\mathbf{a_3}, \mathbf{b_3})$$

$$= \frac{1}{\sqrt{2}} - (-\frac{1}{\sqrt{2}}) + \frac{1}{\sqrt{2}} + \frac{1}{\sqrt{2}} = \frac{4}{\sqrt{2}} = 2\sqrt{2}$$

So, we expect $S = 2\sqrt{2}$ according to the calculations using quantum mechanics.

With n pairs of qubits going through the above process, and n measurements, the set of measurements where A and B happened to use the same basis (i.e. using $\mathbf{a_2}$ and $\mathbf{b_1}$ or using $\mathbf{a_3}$ and $\mathbf{b_2}$, as can be seen from the table above, which implies an efficiency of 2/9) can be used as the shared secret bits (both will have the same outcome, either $+1,+1$ or $-1,-1$), and the proportion of measurements where A and B used different bases can be used to estimate the probabilities $P_{+1,+1}(\mathbf{a_i}, \mathbf{b_j})$,

---

14 Note that with a different entangled pair, i.e. the singlet state, we get $S = -2\sqrt{2}$ in Ekert [1991].

$P_{-1,-1}(\mathbf{a_j}, \mathbf{b_j})$, $P_{+1,-1}(\mathbf{a_j}, \mathbf{b_j})$, and $P_{-1,+1}(\mathbf{a_j}, \mathbf{b_j})$ to be used in calculations of $E(\mathbf{a_1}, \mathbf{b_1})$, $E(\mathbf{a_1}, \mathbf{b_3})$, $E(\mathbf{a_3}, \mathbf{b_1})$, and $E(\mathbf{a_3}, \mathbf{b_3})$.

For example, suppose we want to estimate $E(\mathbf{a_1}, \mathbf{b_1})$, then we need to obtain estimates for $P_{+1,+1}(\mathbf{a_1}, \mathbf{b_1})$, $P_{-1,-1}(\mathbf{a_1}, \mathbf{b_1})$, $P_{+1,-1}(\mathbf{a_1}, \mathbf{b_1})$, and $P_{-1,+1}(\mathbf{a_1}, \mathbf{b_1})$. To estimate $P_{+1,+1}(\mathbf{a_1}, \mathbf{b_1})$, let $N_{\mathbf{a_1},\mathbf{b_1}}$ denote the total number of pairs where A used $\mathbf{a_1}$ and B used $\mathbf{b_1}$, and let $N_{\mathbf{a_1},\mathbf{b_1}}^{+1,+1}$ denote the total number of pairs where A used $\mathbf{a_1}$ and B used $\mathbf{b_1}$ and the outcome is $(+1,+1)$. Then, we compute an estimate $P'_{+1,+1}(\mathbf{a_1}, \mathbf{b_1})$ for $P_{+1,+1}(\mathbf{a_1}, \mathbf{b_1})$:

$$P'_{+1,+1}(\mathbf{a_1}, \mathbf{b_1}) = \frac{N_{\mathbf{a_1},\mathbf{b_1}}^{+1,+1}}{N_{\mathbf{a_1},\mathbf{b_1}}}$$

Note that $N_{\mathbf{a_1},\mathbf{b_1}} = N_{\mathbf{a_1},\mathbf{b_1}}^{+1,+1} + N_{\mathbf{a_1},\mathbf{b_1}}^{-1,-1} + N_{\mathbf{a_1},\mathbf{b_1}}^{+1,-1} + N_{\mathbf{a_1},\mathbf{b_1}}^{-1,+1}$. So, suppose we have 101 pairs (out of $n = 900$ pairs in total) where measurements were done with A using $\mathbf{a_1}$ and B using $\mathbf{b_1}$, and of these, 43 of them resulted in $(+1,+1)$. Then, an estimate for $P_{+1,+1}(\mathbf{a_1}, \mathbf{b_1})$ is $43/101 \approx \frac{1}{2}cos^2(\frac{\pi}{8})$. Similarly, for estimating the other probabilities and then the E values and then S. Note that if there had been "disturbances" so that some third party intercepted the qubits and changed the values or "disturbed" the entanglement, then, the intervention could cause the probability estimates to deviate too much from what is expected, and $S \neq 2\sqrt{2}$. But if indeed S is close enough to the expected value, A and B can be quite certain there had not been any intervention and the secret key they have is secure.

Let's consider an example of an intervention where a third party intercepts each pair of qubits going to A and B (from the source) and always applies the Z basis to "disturb" the qubits – note that an attacker might be a lot more sophisticated than this but this is just an example. This means that with probability $1/2$, the attacker gets $|00\rangle$ (from intercepting/measuring the Bell state) and sends a qubit in state $|0\rangle$ to each, to A and to B. Then, given $|0\rangle = cos(\frac{\pi}{8})|\times\rangle - sin(\frac{\pi}{8})|\div\rangle$, we can write $|00\rangle = |0\rangle \otimes |0\rangle$ as

$$|0\rangle \otimes \left(cos\left(\frac{\pi}{8}\right)|\times\rangle - sin\left(\frac{\pi}{8}\right)|\div\rangle\right) = cos\left(\frac{\pi}{8}\right)|0\rangle|\times\rangle - sin\left(\frac{\pi}{8}\right)|0\rangle|\div\rangle$$

and we can write $|11\rangle = |1\rangle \otimes |1\rangle$ as

$$|1\rangle \otimes \left(sin\left(\frac{\pi}{8}\right)|\times\rangle + cos\left(\frac{\pi}{8}\right)|\div\rangle\right) = sin\left(\frac{\pi}{8}\right)|1\rangle|\times\rangle + cos\left(\frac{\pi}{8}\right)|1\rangle|\div\rangle$$

For each pair of qubits intercepted from the source, where A and B are using bases $\mathbf{a_1}$ and $\mathbf{b_1}$, from the attacker, A and B gets $|00\rangle$ with probability $1/2$ and measures $|0\rangle|\times\rangle$ with probability $cos^2(\frac{\pi}{8})$, i.e. the outcome $(+1,+1)$ is obtained with probability $\frac{1}{2}cos^2(\frac{\pi}{8})$. Similarly, from the above form of $|00\rangle$ the outcome $(+1,-1)$ is obtained with probability $\frac{1}{2}sin^2(\frac{\pi}{8})$. And A and B gets $|11\rangle$ with probability $1/2$ and measures $|1\rangle|\times\rangle$ with probability $sin^2(\frac{\pi}{8})$, i.e. the outcome $(-1,+1)$ is obtained with probability $\frac{1}{2}sin^2(\frac{\pi}{8})$. Similarly, from the above form of $|11\rangle$ the outcome $(-1,-1)$ is obtained with probability $\frac{1}{2}\cos^2(\frac{\pi}{8})$. This seems to agree with what we expect if there was no interception (as in the first line of the table given earlier).

However, now consider each pair of qubits intercepted from the source, where A and B are using bases $\mathbf{a_3}$ and $\mathbf{b_3}$. We can write $|00\rangle = |0\rangle \otimes |0\rangle$ as

$$\frac{1}{\sqrt{2}}(|+\rangle + |-\rangle) \otimes \left(\sin\left(\frac{\pi}{8}\right)|\times'\rangle - \cos\left(\frac{\pi}{8}\right)|\div'\rangle \right)$$

$$= \frac{1}{\sqrt{2}}\sin\left(\frac{\pi}{8}\right)|+\rangle|\times'\rangle + \frac{1}{\sqrt{2}}\sin\left(\frac{\pi}{8}\right)|-\rangle|\times'\rangle - \frac{1}{\sqrt{2}}\cos\left(\frac{\pi}{8}\right)|+\rangle|\div'\rangle$$

$$- \frac{1}{\sqrt{2}}\cos\left(\frac{\pi}{8}\right)|-\rangle|\div'\rangle$$

and we can write $|11\rangle = |1\rangle \otimes |1\rangle$ as

$$\frac{1}{\sqrt{2}}(|+\rangle - |-\rangle) \otimes \left(\cos\left(\frac{\pi}{8}\right)|\times'\rangle + \sin\left(\frac{\pi}{8}\right)|\div'\rangle \right)$$

$$= \frac{1}{\sqrt{2}}\cos\left(\frac{\pi}{8}\right)|+\rangle|\times'\rangle - \frac{1}{\sqrt{2}}\cos\left(\frac{\pi}{8}\right)|-\rangle|\times'\rangle + \frac{1}{\sqrt{2}}\sin\left(\frac{\pi}{8}\right)|+\rangle|\div'\rangle$$

$$- \frac{1}{\sqrt{2}}\sin\left(\frac{\pi}{8}\right)|-\rangle|\div'\rangle$$

Then, from the attacker, A and B gets $|00\rangle$ with probability $1/2$ and gets $|11\rangle$ with probability $1/2$, so that A and B gets $|+\rangle|\times'\rangle$ with probability

$$\frac{1}{2} \cdot \frac{1}{2}\sin^2\left(\frac{\pi}{8}\right) + \frac{1}{2} \cdot \frac{1}{2}\cos^2\left(\frac{\pi}{8}\right) = \frac{1}{4}\left(\sin^2\left(\frac{\pi}{8}\right) + \cos^2\left(\frac{\pi}{8}\right)\right) = \frac{1}{4}$$

i.e. the outcome $(+1,+1)$ is obtained with probability $1/4$. Similarly, the outcomes $(-1,-1)$ is obtained with probability $1/4$, $(+1,-1)$ is obtained with probability $1/4$, and $(-1,+1)$ is obtained with probability $1/4$. This means that with this interception, the statistics to be obtained will likely show that each of the outcomes will be obtained with equal probability (i.e. $1/4$), and we will likely observe that

$$E(\mathbf{a_3}, \mathbf{b_3}) = \frac{1}{4} + \frac{1}{4} - \frac{1}{4} - \frac{1}{4} = 0$$

However, this value of E is different from the last row of the table we saw earlier. The last row is calculated from quantum mechanics as follows. We can write

$$\frac{1}{\sqrt{2}}(|00\rangle + |11\rangle) = \frac{1}{\sqrt{2}}\left(\frac{1}{\sqrt{2}}(|+\rangle + |-\rangle) \otimes \left(\sin\left(\frac{\pi}{8}\right)|\times'\rangle - \cos\left(\frac{\pi}{8}\right)|\div'\rangle \right) \right.$$

$$\left. + \frac{1}{\sqrt{2}}(|+\rangle - |-\rangle) \otimes \left(\cos\left(\frac{\pi}{8}\right)|\times'\rangle + \sin\left(\frac{\pi}{8}\right)|\div'\rangle \right) \right)$$

$$= \frac{1}{\sqrt{2}}\left(\frac{1}{\sqrt{2}}\left(\sin\left(\frac{\pi}{8}\right) + \cos\left(\frac{\pi}{8}\right)\right)|+\rangle|\times'\rangle \right.$$

$$-\frac{1}{\sqrt{2}}\left(\cos\left(\frac{\pi}{8}\right)-\sin\left(\frac{\pi}{8}\right)\right)|-\rangle\,|\times'\rangle$$

$$-\frac{1}{\sqrt{2}}\left(\cos\left(\frac{\pi}{8}\right)-\sin\left(\frac{\pi}{8}\right)\right)|+\rangle\,|\div'\rangle$$

$$-\frac{1}{\sqrt{2}}\left(\cos\left(\frac{\pi}{8}\right)+\sin\left(\frac{\pi}{8}\right)\right)|-\rangle\,|\div'\rangle\Big)$$

$$=\frac{1}{\sqrt{2}}\left(\cos\left(\frac{\pi}{8}\right)|+\rangle\,|\times'\rangle-\sin\left(\frac{\pi}{8}\right)|-\rangle\,|\times'\rangle\right.$$

$$\left.-\sin\left(\frac{\pi}{8}\right)|+\rangle\,|\div'\rangle-\cos\left(\frac{\pi}{8}\right)|-\rangle\,|\div'\rangle\right)$$

The last step uses the identities in the **Aside** earlier. The above calculation suggests that the outcome $(+1,+1)$ should be obtained with probability $\frac{1}{2}\cos^2(\frac{\pi}{8})$, the outcomes $(-1,-1)$ should be obtained with probability $\frac{1}{2}\cos^2(\frac{\pi}{8})$, $(+1,-1)$ should be obtained with probability $\frac{1}{2}\sin^2(\frac{\pi}{8})$, and $(-1,+1)$ should be obtained with probability $\frac{1}{2}\sin^2(\frac{\pi}{8})$. This means that

$$E(\mathbf{a_3}, \mathbf{b_3}) = \frac{1}{2}\cos^2\left(\frac{\pi}{8}\right) + \frac{1}{2}\cos^2\left(\frac{\pi}{8}\right) - \frac{1}{2}\sin^2\left(\frac{\pi}{8}\right) - \frac{1}{2}\sin^2\left(\frac{\pi}{8}\right)$$

$$= \cos^2\left(\frac{\pi}{8}\right) - \sin^2\left(\frac{\pi}{8}\right) = \cos\left(2 \cdot \frac{\pi}{8}\right) = \frac{1}{\sqrt{2}}$$

And so, such an attack can be detected from statistics on pairs where A and B are using bases $\mathbf{a_3}$ and $\mathbf{b_3}$! This is a clever use of all the measurement outcomes – the outcomes where A and B used the same basis will form the shared secret key (since both parties will get the same results), and the outcomes where A and B used different bases will provide statistics for detecting attackers.

There have been multiple demonstrations of QKD protocols in research (e.g. [Chen et al., 2021]) and commercial products for QKD.[15] There also has been a lot of work on conference key agreement, or with $N > 2$ parties wanting to have a common secret key [Murta et al., 2020].

Integrating QKD with today's Internet has been discussed in depth by Mehic et al. [2022]. A REST-based API to allow applications to make method calls to a QKD network to obtain keys is specified as an ETSI (European

---

15 For example, see https://www.idquantique.com/quantum-safe-security/products/cerberis-xg-qkd-system and https://www.global.toshiba/ww/products-solutions/security-ict/qkd.html [last accessed: 15/6/2022].

Telecommunications Standards Institute) standard.[16] A QKD Network Simulator allows further explorations.[17]

3.2.4 Quantum Anonymous Broadcasting

We consider the problem of how anonymous broadcasting can be implemented, i.e. given a collection of n parties (e.g. users or machines), how any one party can broadcast a message (i.e. a one-bit message) to all the other parties, without the other parties knowing who transmitted the message.

The original (non-quantum) form of this problem (and solution) was given as follows, called the Dining Cryptographers Problem:

> Three cryptographers are sitting down to dinner at their favorite three-star restaurant. Their waiter informs them that arrangements have been made with the maître d'hôtel for the bill to be paid anonymously. One of the cryptographers might be paying for dinner, or it might have been NSA (U.S. National Security Agency). The three cryptographers respect each others' right to make an anonymous payment, but they wonder if NSA is paying, They resolve their uncertainty fairly by carrying out the following protocol: Each cryptographer flips an unbiased coin behind his menu, between him and the cryptographer on his right, so that only the two of them can see the outcome. Each cryptographer then states aloud whether the two coins he can see – the one he flipped and the one his left-hand neighbor flipped – fell on the same side or on different sides. If one of the cryptographers is the payer, he states the opposite of what he sees. An odd number of differences uttered at the table indicates that a cryptographer is paying an even number indicates that NSA is paying (assuming that the dinner was paid for only once). Yet if a cryptographer is paying, neither of the other two learns anything from the utterances about which cryptographer it is. [Chaum, 1988, p. 65]

Note that there are three coin flips and three outcomes. NSA paid means none of the cryptographers paid. This protocol is unconditionally secure because suppose one of the nonpayer cryptographer is trying to find out who paid (and assuming one of the cryptographers paid). Then, for this cryptographer (called A), there are two cases:

---

16 See https://www.etsi.org/deliver/etsi_gs/QKD/001_099/014/01.01.01_60/gs_
qkd014v010101p.pdf and https://www.etsi.org/deliver/etsi_gs/QKD/001_099/014/01.01.01_60/
gs_qkd014v010101p.pdf, and also the main site: https://www.etsi.org/committee/1430-qkd [last
accessed: 14/5/2023].
17 See https://open-qkd.eu [last accessed: 14/5/2023].

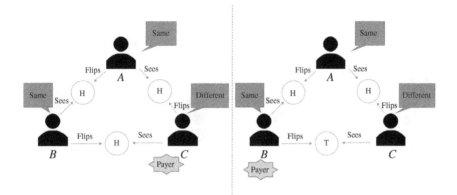

Figure 3.7 Illustration of a scenario where A sees that both the coins (the one A sees and the one A flips) are the same (e.g. both Heads) and B says "same" and C says "different." On the left, if the coin hidden from A (the one between B and C) is Head (same as the other two outcomes), then the one who said "different" is the payer, i.e. C, since C states the opposite of what C says. But on the right, if the coin hidden from A (the one between B and C) is Tail (different from the other two outcomes), then the one who said "same" is the payer, i.e. B, since B states the opposite of what B says. Comparing these two scenarios, the probability of the coin hidden from A (the one between B and C) being Head (or Tail) is 1/2 and so, A cannot know just from one saying "same" and the other saying "different" who the payer might be. Similarly, when C says "same" and B says "different."

1. *The two coins A sees are the same*: One of the other cryptographers will have to say "different" and the other would say "same" (if both say "same," then NSA paid, which is what we assumed is not the case). The one who said "different" is the payer if the outcome hidden from A is the same as the other two outcomes, while the one who said "same" is the payer if the outcome hidden from A is different from the other two outcomes. But the outcome hidden from A has equal probability of being the same or different from the two other outcomes, given a fair coin flip, and so, A cannot tell who the payer is. This is illustrated in Figure 3.7.
2. *The two coins A sees are different*: If the two other cryptographers said "different," then the payer is the one seeing the coin that is the same as the coin hidden from A, as illustrated in Figure 3.8. If the two other cryptographers said "same," then the payer is the one seeing the coin that is different from the coin hidden from A.

We can formalize the reasoning above as follows. Each cryptographer (or party) above is a node in a graph, and the outcome of the coin flipped between a pair of parties is represented by an edge, as illustrated in Figure 3.9, with five nodes and five edges, with each edge representing a bit of information b_i (with value 0 or 1) shared between a pair of parties. Then, whether a party sees both coins (or edges)

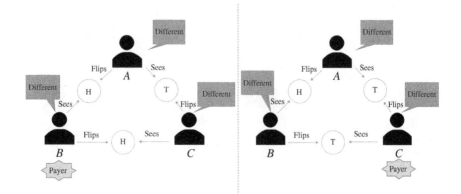

Figure 3.8 Illustration of a scenario where *A* sees that the coins (the one *A* sees and the one *A* flips) are different and both the other two says "different." On the left, if the coin hidden from *A* (the one between *B* and *C*) is Head, then the one who said "different" but the two coins seen are actually the same is the payer, i.e. *B*, since *B* states the opposite of what *B* sees. On the right, if the coin hidden from *A* (the one between *B* and *C*) is Tail, then the one who said "different" (but the two coins seen are actually the same) is the payer, i.e. *C*, since *C* states the opposite of what *C* says. Comparing these two scenarios, the probability of the coin hidden from *A* (the one between *B* and *C*) being Head (or Tail) is 1/2 and so *A* cannot know just from both saying "different" who the payer might be. Similarly when *B* and *C* both says "same."

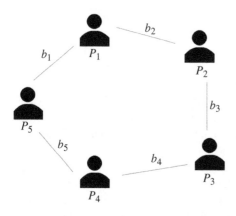

Figure 3.9 Illustration of a graphical representation of the Dining Cryptographers Problem, with five parties and five edges, each edge representing a secret bit shared between the corresponding pair, secret since all the other parties cannot know what this bit is.

same (denoted by 0) or different (denoted by 1), is given by the sum modulo two between adjacent values of b_i, i.e. $b_i \oplus b_{(i+1) \bmod 5}$ for $1 \leq i \leq 5$, that is,

$$P_1 : b_1 \oplus b_2, P_2 : b_2 \oplus b_3, P_3 : b_3 \oplus b_4, P_4 : b_4 \oplus b_5, \text{ and } P_5 : b_5 \oplus b_1.$$

Note that a secret bit b_i between a pair of parties can be obtained via measurement on a Bell pair, shared between the parties in that pair: $\frac{1}{\sqrt{2}}(|00\rangle + |11\rangle)$. More generally, the secret bits shared between pairs of parties are effectively pairwise shared secret keys which can be obtained via a protocol such as QKD.

Based on the presentation in Huang et al. [2022], we can now provide a protocol for *anonymous broadcasting*, which generalizes the solution protocol above for the Dining cryptographers problem for n parties. Let us assume they are organized in a ring as shown in Figure 3.9 – we then generalize this to a fully connected graph. Suppose one party (call this the speaker) wants to broadcast one bit of information to the rest of the $n - 1$ parties without revealing its identity, the protocol is as follows.

For each round:

1. If the speaker (say party indexed by s) wishes to broadcast a value $c = 0$, then the speaker simply sets $p_s = b_s \oplus b_{(s+1) \bmod n}$ as the value to announce; if the speaker wishes to broadcast a value $c = 1$, then the speaker flips the value of one of b_s or $b_{(s+1) \bmod n}$. Let us say the speaker flips b_s. Then the speaker sets $p_s = \overline{b_s} \oplus b_{(s+1) \bmod n}$ as the value to announce; all the others (for $i \neq s$) are non-speakers and just set $p_i = b_i \oplus b_{(i+1) \bmod n}$ as the value to announce.
2. Each party announces (via some classical broadcast channel, not anonymous) its p_i value to everyone else, and then since each party has everyone's p_i values, each can compute the overall parity value: $p = p_1 \oplus \cdots \oplus p_n$.

Then, note that if $c = 0$, that is, the speaker is broadcasting 0, then

$$
\begin{aligned}
p &= p_1 \oplus \cdots \oplus p_n \\
&= (b_1 \oplus b_2) \oplus (b_2 \oplus b_3) \oplus \cdots \oplus (b_{n-2} \oplus b_{n-1}) \oplus (b_{n-1} \oplus b_n) \oplus (b_n \oplus b_1) \\
&= (b_1 \oplus b_1) \oplus (b_2 \oplus b_2) \oplus \cdots \oplus (b_n \oplus b_n) = 0
\end{aligned}
$$

since each bit b_i occurs exactly twice in the expression. But if $c = 1$, that is, the speaker is broadcasting 1, then

$$
\begin{aligned}
p &= p_1 \oplus \cdots \oplus p_{s-1} \oplus p_s \oplus \cdots \oplus p_n \\
&= (b_1 \oplus b_2) \oplus \cdots \oplus (b_{s-1} \oplus b_s) \oplus (\overline{b_s} \oplus b_{s+1}) \oplus \cdots \oplus (b_n \oplus b_1) \\
&= (b_1 \oplus b_1) \oplus (b_2 \oplus b_2) \oplus \cdots \oplus (b_s \oplus \overline{b_s}) \oplus \cdots \oplus (b_n \oplus b_n) = 1
\end{aligned}
$$

Since p is computed by everyone, everyone receives the bit broadcasted by the speaker, but no one knows who the speaker is, since each b_i has probability of 1/2 of being 0 or 1. Note that the quantum network here is used only to create the pairwise secret bits.

A simple extension of the above can be used for a (general) graph (rather than just the ring topology) where secret bits are shared among all pairs of users, as shown in Figure 3.10.

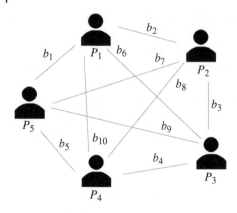

Figure 3.10 Illustration of a graphical representation of the Dining Cryptographers Problem, with 5 parties and 10 edges (fully connected), each edge representing a secret bit shared between the corresponding pair, secret since all the other parties cannot know what this bit is.

The protocol is then as follows. For each round:

1. If the speaker (say party indexed by s) wishes to broadcast a value $c = 0$, then the speaker simply sets $p_s = \oplus_{i \in B_s} b_i$, where B_s is the set of all edges (bits) incident at P_s as the value to announce; if the speaker wishes to broadcast a value $c = 1$, then the speaker flips the value of one of $b_i \in B_s$, say b_s. That is, the speaker sets $p_s = \bar{b}_s \oplus (\oplus_{i \in B_s - \{b_s\}} b_i)$ as the value to announce; all the others (for $j \neq s$) are non-speakers and just set $p_j = \oplus_{i \in B_j} b_i$ as the value to announce.
2. Each party announces its p_i value to everyone else, and then since each party has everyone's p_i values, each can compute the overall parity value: $p = p_1 \oplus \cdots \oplus p_n$.

We can use the scenario in Figure 3.10 to illustrate this version of the protocol. Suppose the speaker is P_2, then the speaker announces: $p_2 = \bar{b}_2 \oplus b_7 \oplus b_8 \oplus b_3$, and the rest announces: $p_1 = b_1 \oplus b_{10} \oplus b_6 \oplus b_2$, $p_3 = b_3 \oplus b_6 \oplus b_9 \oplus b_4$, $p_4 = b_4 \oplus b_8 \oplus b_{10} \oplus b_5$, and $p_5 = b_5 \oplus b_9 \oplus b_7 \oplus b_1$. When a speaker broadcasts $c = 1$, the overall parity p is then:

$$
\begin{aligned}
p &= p_1 \oplus p_2 \oplus p_3 \oplus p_4 \oplus p_5 \\
&= (b_1 \oplus b_{10} \oplus b_6 \oplus b_2) \oplus (\bar{b}_2 \oplus b_7 \oplus b_8 \oplus b_3) \\
&\quad \oplus (b_3 \oplus b_6 \oplus b_9 \oplus b_4) \oplus (b_4 \oplus b_8 \oplus b_{10} \oplus b_5) \\
&\quad \oplus (b_5 \oplus b_9 \oplus b_7 \oplus b_1) \\
&= (b_1 \oplus b_1) \oplus (b_2 \oplus \bar{b}_2) \oplus (b_3 \oplus b_3) \oplus (b_4 \oplus b_4) \oplus (b_5 \oplus b_5) \\
&\quad \oplus (b_6 \oplus b_6) \oplus (b_7 \oplus b_7) \oplus (b_8 \oplus b_8) \oplus (b_9 \oplus b_9) \oplus (b_{10} \oplus b_{10}) = 1
\end{aligned}
$$

and $p = 0$ when $c = 0$.

Note that the pairwise secret bits are used in each round, and so, if the protocol needs to proceed for multiple rounds (e.g. the speaker has a lot more than one bit to say!), then between each pair, multiple secret bits are required (secret in that no one else except the pair knows) – the QKD protocol can be used to create such a string of r secret bits between each pair, for r rounds.

As shown in Huang et al. [2022] and based on Broadbent and Tapp [2007], ingenious variations and extensions of the ideas lead to other interesting protocols such as *veto*, where if one or more users can veto a proposal in such a way that all those who vetoed have their identities hidden, *notification*, where a user can notify one or more other users with the identities of the notifier and recipients (and number of recipients) remaining hidden, *collision detection*, where more than one speaker in an anonymous broadcast protocol can be detected, and *anonymous private message transmission*, where a sender can anonymously transmit an m-bit message to a receiver (receiver does not know who the sender is).

Another quantum anonymous transmission protocol is by Christandl and Wehner [2005] using generalized GHZ states as follows.

1. All n-party share the n-qubit state $\frac{1}{\sqrt{2}}(|0\rangle^{\otimes n} + |1\rangle^{\otimes n})$.
2. The speaker applies a phase flip Z gate to its qubit (i.e. to its part of the shared state) if $c = 1$ (i.e. it wants to broadcast a 1 anonymously) and does nothing otherwise.
3. Each party (including the speaker) does the following:
   (i) applies the Hadamard transform H to its qubit,
   (ii) measures its qubit in the computational basis,
   (iii) broadcasts its measurement result,
   (iv) receives the announced measurement outcomes from everyone and counts the total number of 1s (call this number k), and
   (v) if k is odd, it concludes $c = 1$, otherwise it concludes $c = 0$
4. The protocol aborts if one or more players do not use the broadcast channel.

The anonymity in this protocol is due to the fact that in the n-qubit state $\frac{1}{\sqrt{2}}(|0\rangle^{\otimes n} + |1\rangle^{\otimes n})$ shared among n parties, if any one party applies the Z gate to its qubit, the state obtained is $\frac{1}{\sqrt{2}}(|0\rangle^{\otimes n} - |1\rangle^{\otimes n})$. And there is no way that one can trace back, just from this state, who applied the Z to a qubit. Note that we want $\frac{1}{\sqrt{2}}(|0\rangle^{\otimes n} + |1\rangle^{\otimes n})$ when $c = 0$ and $\frac{1}{\sqrt{2}}(|0\rangle^{\otimes n} - |1\rangle^{\otimes n})$ when $c = 1$, i.e. we want: $\frac{1}{\sqrt{2}}(|0\rangle^{\otimes n} + (-1)^c|1\rangle^{\otimes n})$. Also, step 3 in the protocol is necessary to determine if there was really a change in the phase of the global state, that is, if anyone did apply a Z gate to a qubit in the shared state. To see why, we consider the calculations as follows, following step 3(i):

$$H^{\otimes n}\left(\frac{1}{\sqrt{2}}(|0\rangle^{\otimes n} + (-1)^c|1\rangle^{\otimes n})\right)$$
$$= \frac{1}{\sqrt{2}}(H^{\otimes n}|0\rangle^{\otimes n} + (-1)^c H^{\otimes n}|1\rangle^{\otimes n})$$

$$= \frac{1}{\sqrt{2}}((H\,|0\rangle)^{\otimes n} + (-1)^c(H\,|1\rangle)^{\otimes n})$$

$$= \frac{1}{\sqrt{2}}\left(\left(\frac{1}{\sqrt{2}}(|0\rangle + |1\rangle)\right)^{\otimes n} + (-1)^c\left(\frac{1}{\sqrt{2}}(|0\rangle - |1\rangle)\right)^{\otimes n}\right)$$

$$= \frac{1}{\sqrt{2}}\left(\left(\frac{1}{\sqrt{2}}\right)^n(|0\rangle + |1\rangle)^{\otimes n} + (-1)^c\left(\frac{1}{\sqrt{2}}\right)^n(|0\rangle - |1\rangle)^{\otimes n}\right)$$

$$= \left(\frac{1}{\sqrt{2}}\right)^{n+1}((|0\rangle + |1\rangle)^{\otimes n} + (-1)^c(|0\rangle - |1\rangle)^{\otimes n})$$

$$= \left(\frac{1}{\sqrt{2}}\right)^{n+1}\left(\sum_{x\in\{0,1\}^n} |x\rangle + (-1)^c \sum_{x\in\{0,1\}^n} (-1)^{[x]}\,|x\rangle\right)$$

$$= \left(\frac{1}{\sqrt{2}}\right)^{n+1}\sum_{x\in\{0,1\}^n} (1 + (-1)^{c+[x]})\,|x\rangle$$

where $[x]$ is used to denote the Hamming weight of the string x, i.e. in this case, the number of 1s in x. With the expression above, if $c = 0$, then

$$1 + (-1)^{c+[x]} = 1 + (-1)^{0+[x]} = \begin{cases} 0, & \text{if } [x] \text{ is odd} \\ 2, & \text{if } [x] \text{ is even} \end{cases}$$

so that the above reduces to:

$$\left(\frac{1}{\sqrt{2}}\right)^{n+1}\sum_{\substack{x\in\{0,1\}^n \\ \text{such that } [x] \text{ is even}}} |x\rangle$$

that is, the amplitudes are only on the states with $[x]$ even. But if $c = 1$, then

$$1 + (-1)^{c+[x]} = 1 + (-1)^{1+[x]} = \begin{cases} 2, & \text{if } [x] \text{ is odd} \\ 0, & \text{if } [x] \text{ is even} \end{cases}$$

so that the above reduces to:

$$\left(\frac{1}{\sqrt{2}}\right)^{n+1}\sum_{\substack{x\in\{0,1\}^n \\ \text{such that } [x] \text{ is odd}}} |x\rangle$$

that is, the amplitudes are only on the states with $[x]$ odd. Then, upon measurement in step 3(ii), we have a certain value of x, say x' corresponding to the state $|x'\rangle$ obtained, and $k = [x']$. From the above, if $c = 0$, given the amplitudes are only on states with even Hamming weights, k has to be even. And similarly, if $c = 1$, k is odd, which is what we see in step 3(v). We assume that there is only one speaker in each round – there are protocols to deal with multiple senders/speakers and to send a qubit rather than a classical bit [Christandl and Wehner, 2005].

3.2.5 Quantum Voting

Voting is a key concept in democratic societies and quantum electronic voting protocols have been extensively discussed in Arapinis et al. [2021]. Here, we outline the voting scheme using quantum information from Khabiboulline et al. [2021].

The general idea is to use a quantum state as a kind of "ballot paper" where a voter can register a vote (a vote here is binary, e.g. either "yes" [say 1] or "no" [say 0]), and the voting result is obtained by counting the number of 1s and 0s obtained. A key consideration is that a voter should not be able to provide multiple votes (undetected), thereby compromising the integrity of the election.

We have a tallyman that samples a random bit string $x \in \{0,1\}^n$, where $n = O(N^2)$ for the protocol to work correctly, and prepares N copies of the ballot state, where N is the number of voters:

$$|\psi_x\rangle = \frac{1}{\sqrt{n}} \sum_{i=0}^{n-1} (-1)^{x_i} |i\rangle$$

where x_i is the ith bit of the n-bit string $x = x_0 \ldots x_{n-1}$, and the ballot state uses $\lceil \log n \rceil$ qubits. For example, for $n = 6$, we use three qubits, and a ballot state with $x = 101100$ is a superposition of six states:

$$\frac{1}{\sqrt{6}}(-|000\rangle + |001\rangle - |010\rangle - |011\rangle + |100\rangle + |101\rangle)$$

Note that the coefficients of $|110\rangle$ and $|111\rangle$ are considered 0 since the string only has 6 bits. The ballot state marks a 1 in the ith position via the negative sign of $|i\rangle$, i.e. each component is $(-1)^{x_i} |i\rangle$.

How does a voter cast a vote? On receiving a ballot state, a voter performs measurements in a basis comprising projectors on superpositions of two states of the form $\frac{1}{2}(|i\rangle + |j\rangle)$ and $\frac{1}{2}(|i\rangle - |j\rangle)$ where $i \neq j$. If we think of the six component states in the ballot state as nodes of a graph, then an edge between two nodes (i, j) can be represented by a term of the form $(|i\rangle + |j\rangle)$ or $(|i\rangle - |j\rangle)$. One example of a measurement basis is the set of projectors (using these states):

$$\left\{ \frac{1}{\sqrt{2}}(|000\rangle + |011\rangle), \frac{1}{\sqrt{2}}(|000\rangle - |011\rangle), \right.$$
$$\frac{1}{\sqrt{2}}(|001\rangle + |100\rangle), \frac{1}{\sqrt{2}}(|001\rangle - |100\rangle),$$
$$\left. \frac{1}{\sqrt{2}}(|010\rangle + |101\rangle), \frac{1}{\sqrt{2}}(|010\rangle - |101\rangle) \right\}$$

which corresponds to three edges between states, and when used for measurement would effectively pick an edge from the three possible edges, yielding either $(|i\rangle + |j\rangle)$ or $(|i\rangle - |j\rangle)$ which corresponds to the parity of x_i and x_j given the definition of the ballot state (since if we get the resulting state $(|i\rangle + |j\rangle)$, then the ballot state contains either the term $|i\rangle + |j\rangle$ or $-|i\rangle - |j\rangle$, so that it must be that $x_i = x_j$, and if we get the resulting state $(|i\rangle - |j\rangle)$, then the ballot state contains either the term $|i\rangle - |j\rangle$ or $-|i\rangle + |j\rangle$, so that it must be that $x_i \neq x_j$). This means that the voter can learn the parity p of x_i and x_j. This parity is then used to encode its vote, i.e. if its vote is $v \in \{0, 1\}$, then its encoded vote is $a = p \oplus v$. For example, rewriting the ballot state example above to:

$$\frac{1}{\sqrt{6}}(-(|000\rangle + |011\rangle)) + (|001\rangle + |100\rangle) - (|010\rangle - |101\rangle))$$

if the projectors above were used for measurement on the ballot state example, the voter might get the state:

$$\frac{1}{\sqrt{2}}(|000\rangle + |011\rangle)$$

and find out the parity between x_0 and x_3, i.e. it finds that $x_0 = x_3$, and use $p = 0$.

Each voter broadcasts its encoded vote a and its edge (i, j), i.e. each vote is accompanied by a distinct edge. Note that other voters, on hearing this, cannot yet decode the vote since they do not know x.

Once all the voters have broadcasted their encoded vote (and the corresponding edge), that is, have effectively cast their votes, the tallyman reveals x to all so that each voter can decipher all the votes from everyone, and each determine the winner.

Only votes accompanied by distinct edges are counted, to prevent double voting. For example, if there are two votes accompanied by the same edge, then only the first one broadcasted can be counted. Or if there are two votes where the two edges share a node, then, only the first one broadcasted is counted. This would prevent, for example, two voters colluding to forge a vote, e.g. one voter used the edge (i, k) and another used the edge (k, j), then because $x_i \oplus x_j = (x_i \oplus x_k) \oplus (x_k \oplus x_j)$, then they can collude to send in a vote accompanied by the edge (i, j) – two voters then provide three votes! The bit string length n can be made large enough so that the probability of two honest voters sampling, by chance, overlapping edges can be minimized.

What if a voter tries to present two votes, one encoded with the parity obtained from the actual edge obtained from measurement on the ballot state, and another encoded by guessing the parity of any other random edge? Note that the probability of guessing the parity of the other edge right is 1/2. Even if the guess was wrong, the additional vote could mess up results. This is prevented by repeating the voting protocol $O(N)$ rounds and averaging the votes over the rounds – this means that the additional votes will on average have an equal number of 0s and 1s, and so, not affect the final result.

The voters broadcast their votes and accompanying edges using an anonymous broadcast protocol so no one can know how each voter voted. A much deeper analysis of the cryptographic properties of the protocol is given in Khabiboulline et al. [2021].

3.2.6 Quantum Byzantine Generals Solution

The *Byzantine Generals Problem* (BGP) is a well-known challenging problem in distributed computing. The problem is about how to achieve consensus among multiple parties (e.g. a node in a distributed system) when some of the parties fail or a party receives conflicting information from different parties.

The *Byzantine Generals Problem (BGP)* or *Byzantine Agreement (BA)* problem can be stated using a metaphor using the idea of the "Byzantine army," as follows:

> We imagine that several divisions of the Byzantine army are camped outside an enemy city, each division commanded by its own general. The generals can communicate with one another only by messenger. After observing the enemy, they must decide upon a common plan of action. However, some of the generals may be traitors, trying to prevent the loyal generals from reaching agreement. The generals must have an algorithm to guarantee that:
>
> A. All loyal generals decide upon the same plan of action. ...
> B. A small number of traitors cannot cause the loyal generals to adopt a bad plan.... [Lamport et al., 1982, p. 382–383]

More specifically:

> A commanding general must send an order to his $n - 1$ lieutenant generals such that
>
> IC1. All loyal lieutenants obey the same order.
> IC2. If the commanding general is loyal, then every local lieutenant obeys the orders he sends.
>
> ...Note that if the commander is loyal, then IC1 follows from IC2. However, the commander need not be loyal. [Lamport et al., 1982, p. 384]

IC stands for interactive consistency. An assumption is that the generals can send messages directly to each other. The commanding general can be disloyal in that it sends different messages to the lieutenant generals, thereby sabotaging the attack.

It was noted that there is no solution for the BA problem unless more than two-thirds of the generals are loyal, but if additional conditions are imposed so that

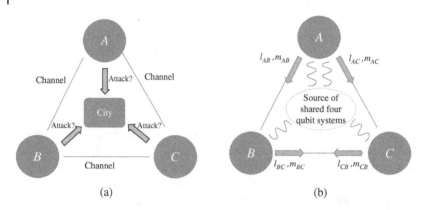

Figure 3.11 (a) An illustration of the metaphor of three generals (with their armies) deciding whether to attack the city together – the idea is that they should agree on a common plan to succeed – all should attack together, or not at all. (b) An illustration of the protocol for DBA with three parties connected by error-free reliable classical channels.

generals use signed messages (and so, cannot forge messages), then the problem can be solved even if less than two-thirds of the generals are loyal [Lamport et al., 1982]. A variation of the BA problem is called the Detectable Byzantine Agreement (DBA) problem, where loyal generals perform the same action or all can abort (if something wrong is detected). The DBA problem can be solved using correlated lists, which each party holds – the list of numbers held by each party has some correlations with the lists of others; we explain the notion of correlated lists later. Shared quantum states are used to create such correlated lists among multiple parties.

We consider a quantum solution to DBA from Gaertner et al. [2008] involving three parties, as illustrated in Figure 3.11.

Each party (i.e. general) has a private list l_A, l_B, and l_C, respectively, with the same length L. The elements of l_A are random trits (values 0, 1, or 2), and the elements of l_B and l_C are random bits (values 0 or 1). A condition is that the three lists, though private, are correlated in this way; at position $1 \leq j \leq L$ in the list, we have one of the following combinations (correlations):

- 000 (i.e. $l_{Aj} = 0$, $l_{Bj} = 0$, and $l_{Cj} = 0$),
- 111 (i.e. $l_{Aj} = 1$, $l_{Bj} = 1$, and $l_{Cj} = 1$), or
- 201 (i.e. $l_{Aj} = 2$, $l_{Bj} = 0$, and $l_{Cj} = 1$) or 210 (i.e. $l_{Aj} = 2$, $l_{Bj} = 1$, and $l_{Cj} = 0$), with equal probability.

So, though each list is private, one party can know something about the other lists via the above correlations. For example, if at position j in A's list is such that $l_{Aj} = 2$, then A knows that at position j in B's and C's lists, either $l_{Bj} = 0$ and $l_{Cj} = 1$

or $l_{Bj} = 1$ and $l_{Cj} = 0$. If $l_{Aj} = 0$, then A knows that $l_{Bj} = 0$ and $l_{Cj} = 0$, and if $l_{Aj} = 1$, then A knows that $l_{Bj} = 1$ and $l_{Cj} = 1$.

In terms of distributed systems, "loyal" could mean working fine, and "disloyal" or being a "traitor" could mean there is some node failure or malfunction, or a node sending inconsistent data (explained below), or a node trying to make the generals do different actions.

The protocol proceeds as follows, with A as the commanding general:

Step 1 (quantum part): All parties generate the lists l_A, l_B, and l_C using quantum information with the correlations as mentioned earlier. We explain in detail how this is obtained later. Otherwise, they agree to abort the protocol.

Step 2 (classical part): The parties transmit the following messages.

A sends a message m_{AB} to B which is 0 or 1 (corresponding to say "abort" or "attack"), and also sends a message $m_{AC} = m_{AB}$ to C. As a way to show that the message m_{AB} is indeed from A, this message is accompanied by a list l_{AB} $(=l_{AC})$, which contains the positions in l_A with value the same as m_{AB}. For example, if $m_{AB} = 0$ and A's list is $\{2, 0, 0, 2, 1, 1, 0, 0, 2, ...\}$, then l_{AB} is $\{2, 3, 7, 8, ...\}$. Similarly, a loyal A also sends the list $l_{AC} = l_{AB}$ to C. We discuss what happens when B receives the message from A (but the discussion applies to C as well due to symmetry).

When B receives m_{AB} and l_{AB}, there are two possibilities:

(i) if l_{AB} is the right length (which should be roughly $L/3$ given that A's list is a random list of trits), and m_{AB}, l_{AB}, and l_B satisfy the correlations above (i.e. m_{AB}, l_{AB}, and l_B are said to be *consistent*), then B will follow the plan m_{AB}, unless C convinces it that A is a traitor (as we see later).

(ii) If m_{AB}, l_{AB}, and l_B are *inconsistent*, then B concludes that A is a traitor, and will not follow m_{AB} and its plan depends on it having an agreement with C as we see later (see comment marked (*)). For example, if $m_{AB} = 0$ and l_{AB} from A is $\{2, 5, 6, 7, ...\}$, and B's list is $l_B = \{1, 0, 0, 0, 1, 1, 0, 0, 0, ...\}$, then B concludes that the data from A is inconsistent since at positions 5 and 6, $l_{B5} = l_{B6} = 1$, and due to the correlations above, we would be expecting $l_{A5} = 1$ or $l_{A5} = 2$, and $l_{A6} = 1$ or $l_{A6} = 2$, that is, the value in A's list at positions 5 and 6 should not be 0.

B sends a message m_{BC}, accompanied by list l_{BC}, to C as well; m_{BC} can be 0, 1 or $\perp$ (meaning "I have inconsistent data"). Note that, B's message to C should be what it received from A, i.e. $m_{BC} = m_{AB}$ and $l_{BC} = l_{AB}$.

When C receives m_{BC} and l_{BC} from B, it already has m_{AC} and l_{AC} from A, there are six possibilities:

(i) m_{AC}, l_{AC}, and l_C are consistent, and m_{BC}, l_{BC}, and l_C are consistent, and $m_{AC} = m_{BC}$: C follows the plan $m_{AC} = m_{BC}$. Here, when A follows its own plan that it sent to B and C, and from what B does above, all A, B, and C would then be following the same plan!

(ii) $m_{AC}, l_{AC},$ and l_C are consistent, and $m_{BC}, l_{BC},$ and l_C are consistent, and $m_{AC} \neq m_{BC}$: C then knows that A is the traitor since only A can send consistent data to both B and C (B cannot manufacture the consistent data here since it does not have l_A), and assuming only one traitor, i.e. B is loyal. By symmetry, B also concludes that A is the traitor and C is loyal, and C realizes this. Then, C and B will follow a previously decided plan, e.g. 0.

(iii) $m_{AC}, l_{AC},$ and l_C are consistent, and $m_{BC} = \perp$: C will follow m_{AC}. It is interesting to note here that there is no way for C to know if B is lying or not (that B actually received inconsistent data from A, or that B received consistent data from A but sending $\perp$ to C). Suppose B is lying (and is a traitor), (and A is loyal), then C by following m_{AC} fulfills its condition (IC2) of doing what A commands. Suppose B is not lying, then A is a traitor, then C by following m_{AC} fulfills its condition (IC1) since B would do the same ((*): B would do the same since it received from C the data $m_{CB} = m_{AC}$ and $l_{CB} = l_{AC}$ which is consistent and received inconsistent data from A, then B will follow the plan from $C, m_{CB} = m_{AC}$, i.e. B and C would take the same action m_{AC}, satisfying (IC1)).

(iv) $m_{AC}, l_{AC},$ and l_C are consistent, and $m_{BC}, l_{BC},$ and l_C are inconsistent: C then determines that B is the traitor and A is loyal, and follows m_{AC}.

(v) $m_{AC}, l_{AC},$ and l_C are inconsistent, and $m_{BC}, l_{BC},$ and l_C are consistent: C determines that A is the traitor and then will follow m_{BC}.

(vi) $m_{AC}, l_{AC},$ and l_C are inconsistent, and $m_{BC} = \perp$: both C and B now know that A is the traitor and will follow a previously decided plan, e.g. 0.

(Note we don't have the case $m_{AC}, l_{AC},$ and l_C are inconsistent, and $m_{BC}, l_{BC},$ and l_C are inconsistent, since A is the traitor and B is loyal and being loyal, B when it receives inconsistent data from A will forward $\perp$ to C (which is case (vi)), and B when it receives consistent data from A will forward the consistent data to C (which is case (v))).

With the above protocol, we see that the conditions of DBA are satisfied.

The quantum part of the protocol is step 1, which can be achieved by a source (as illustrated in Figure 3.11) emitting a large number of four-qubit systems $abcd$ in the four-qubit state:

$$|\psi\rangle_{abcd} = \frac{1}{2\sqrt{3}}(2\,|0011\rangle - |0101\rangle - |0110\rangle - |1001\rangle - |1010\rangle + 2\,|1100\rangle)_{abcd}$$

For each such system, two qubits (a) and (b) are sent to A and one qubit (c) to B and one qubit (d) to C. Suitable measurements on these qubits by A, B, and C then produces the lists l_A, l_B, and l_C with the correlations as above. Note that, for each four-qubit system, if all parties measured in the same basis, one of the following states among $|0011\rangle, |0101\rangle, |0110\rangle, |1001\rangle, |1010\rangle, |1100\rangle$ will be obtained:

- 1100, that is, A obtains 11 (both its qubits are $|1\rangle$), B obtains 0 and C obtains 0, in which A treats 11 as a "0," and so, we can get the combination 000, or
- 0011, that is, A obtains 00 (both its qubits are $|0\rangle$), B obtains 1 and C obtains 1, in which A treats 00 as a "1," and so, we can get the combination 111, or
- 0101, 0110, 1001, or 1010, in which A obtains 01 or 10, either of which is treated as "2," B and C obtains 01 or 10, and so, we can get the combination 210 or 201.

Such measurements done for a series of four-qubit systems received by A, B, and C then results in the lists l_A, l_B, and l_C with the required correlations. Note that the state $|\psi\rangle_{abcd}$ has an interesting property that it remains invariant under the same unitary transformation applied to the four qubits [Gaertner et al., 2008], that is: $U \otimes U \otimes U \otimes U |\psi\rangle_{abcd} = |\psi\rangle_{abcd}$. This means that all parties just need to keep bits resulting from measurements in the same basis, whatever that may be.

Then, another series of tests are carried out as follows to ensure the correlations among the generated lists. C randomly chooses a position k_C in its list l_C and asks A and B, via the pairwise channels, of their values in their respective lists at the same position – if all parties did measure in the same basis, then, the entries at position k_C from the three lists should be correctly correlated. Such test entries in the lists are discarded to maintain the secrecy condition. Then, the parties exchange roles with B choosing a new position k_B and asking A and C for their corresponding entries and checking the correlation, and then A's turn with position k_A and so on, and the process repeats, until a large number of such tests have been carried out.

At the end, the observed quantum error ratio (QER), i.e. the ratio of errors over the total number of four-qubit system detection events is computed – each loyal general then decides to abort or use the generated lists based on some threshold QER.

The following table shows examples of entries of lists obtained experimentally as reported in Gaertner et al. [2008].

Position	l_A	l_B	l_C
1	2	1	0
2	0	0	0
3	0	0	0
4	2	0	1
5	1	1	1
6	1	1	1
7	0	0	0
8	0	0	0
⋮	⋮	⋮	⋮

Other solutions have been developed for the DBA problem, including generalizations to any number of dishonest parties and with more than three generals as in Cholvi [2022], and earlier techniques such as Ben-Or and Hassidim [2005] and one using QKD protocols [Iblisdir and Gisin, 2004], to mention only a few.

3.2.7 Quantum Secret Sharing

We consider the problem of sharing a secret among multiple parties so that each party has part of the secret and cannot use the part to obtain the full secret, and the secret can only be put together by combining the parts from multiple (if not all) parties. More generally, one can consider a secret sharing scheme called the (k, n)-threshold scheme as follows: a secret is divided into n shares such that any $k \leq n$ of those shares can be used to assemble the secret, but any $k - 1$ shares would not provide any information about the secret.

In the classical case, consider three parties A, B, and C. A wants to share a secret s, an n-bit string, shared among B and C so that only when B and C get together can the secret be reconstructed. One way to do this is for A to take the secret s and an n-bit random string r and compute $s' = s \oplus r$ (where "$\oplus$" here denotes the bitwise addition modulo 2). A then gives s' to B and the string r to C. Note that at this point, neither B or C knows anything about s, but when they get together, they compute $s' \oplus r = s \oplus r \oplus r = s$, and so, obtains the secret s. There are other protocols to split s into more than two shares. Such a protocol is subject to eavesdropping so that the shares r and s' might need to be encrypted using keys shared between A and B and between A and C. A protocol such as QKD as we have seen can be used to general such keys for secret sharing.

Alternatively, quantum approaches using GHZ states can be used to split the secret as well as detect eavesdropping, without using QKD, and has been proposed as far back as the late 1990s [Hillery et al., 1999].

Here, we outline briefly the idea of a quantum secret sharing scheme, i.e. a $(2,3)$-threshold scheme, where a secret in this case is a quantum state of a quantum trit (or qutrit), as described in Cleve et al. [1999].

The secret qutrit $\alpha |0\rangle + \beta |1\rangle + \gamma |2\rangle$, with $\alpha^2 + \beta^2 + \gamma^2 = 1$, is encoded using three qutrits as follows:

$$\frac{1}{\sqrt{3}} \left[\alpha(|000\rangle + |111\rangle + |222\rangle) + \beta(|012\rangle + |120\rangle + |201\rangle) + \gamma(|021\rangle \right.$$
$$\left. + |102\rangle + |210\rangle) \right]$$

And then each qutrit is one of three shares. From any one share, no information about the secret can be obtained since the party with the one share has equal mix of $|0\rangle$, $|1\rangle$ and $|2\rangle$ – for example, consider the party with the first qutrit, and suppose measurements have been made on the other two qutrits but not told to this party. One can see that without knowledge of the other two qutrits, the possible states, i.e. the other two qutrits, can be any of the nine combinations $|00\rangle$, $|01\rangle$, $|02\rangle$, $|10\rangle$, $|11\rangle$, $|12\rangle$, $|20\rangle$, $|21\rangle$, and $|22\rangle$, and this party has equal probability (i.e. $(\alpha^2 + \beta^2 + \gamma^2)/3$) of its qutrit being measured to be $|0\rangle$, $|1\rangle$, or $|2\rangle$.

But now, suppose someone has any two of the three shares, say the last two qutrits. Then, the above encoding can be rewritten as follows:

$$\alpha(|000\rangle + |111\rangle + |222\rangle) + \beta(|012\rangle + |120\rangle + |201\rangle) + \gamma(|021\rangle + |102\rangle + |210\rangle)$$

$$= \alpha|000\rangle + \beta|012\rangle + \gamma|021\rangle + \alpha|111\rangle + \beta|120\rangle + \gamma|102\rangle + \alpha|222\rangle$$
$$+ \beta|201\rangle + \gamma|210\rangle$$

$$= |0\rangle \oplus (\alpha|00\rangle + \beta|12\rangle + \gamma|21\rangle) + |1\rangle \oplus (\alpha|11\rangle + \beta|20\rangle + \gamma|02\rangle)$$
$$+ |2\rangle \oplus (\alpha|22\rangle + \beta|01\rangle + \gamma|10\rangle)$$

Then, whatever the first qutrit would be, this someone then practically has access to the secret qutrit in the sense of the amplitudes for three distinct states, e.g. if the first qutrit was $|0\rangle$, then by treating the remaining two qutrits $|00\rangle$ as "$|0\rangle$," $|12\rangle$ as "$|1\rangle$," and $|21\rangle$ as "$|2\rangle$," then this someone effectively has the secret qutrit: α "$|0\rangle$" + β "$|1\rangle$" + γ "$|2\rangle$." Similarly, if the first qutrit was $|1\rangle$ or $|2\rangle$, but a different mapping. Since the remaining two qutrits are different in different cases of the first qutrit, suppose the first qutrit was measured to be $|1\rangle$, if this someone measures and gets $|20\rangle$ say, then this someone also knows the first qutrit was $|1\rangle$.

We have only scratched the surface – a much deeper analysis and theorems on the theoretical properties of the secret sharing schemes are provided in Cleve et al. [1999].

There are still many more recent developments in the area of secret sharing, e.g. the (3,5)-threshold scheme in Long et al. [2021] and see others Tsai et al. [2022] and Zhang et al. [2019] to name just several.

3.2.8 Quantum Oblivious Transfer (OT)

OT is the problem of allowing data transfer or answering of queries in such a way that certain secrecy or privacy properties are preserved. More specifically, suppose we have two parties, A and B, and A has a database with two items d_0 and d_1 (each of which are n-bit strings, say) and (i) B would like to retrieve one of the two items but does not want A to know which one it retrieved, preserving its privacy and (ii) A can be assured that B does not get both items. The particular case with A having two items is called 1-2 OT. A trivial solution for (i) is for A to just send all (or both) its items (or its whole database) to B, but this would not satisfy (ii).

We consider the 1-2 OT protocol, using the ideas from the one presented by Vidick and Wehner.[18] Below A has the items $d_0, d_1 \in \{0,1\}^n$, and B is requesting an item $y \in \{0,1\}$. The aim is for the protocol to compute $f_B((d_0, d_1), y) = d_y$ for B and $f_A((d_0, d_1), y) = \perp$ for A.

1. A selects a random $2n$-bit string $x = x_1 \ldots x_{2n} \in \{0,1\}^{2n}$ and a random basis string $\theta = \theta_1 \ldots \theta_{2n} \in \{0,1\}^{2n}$, sends to B bits encoded in the chosen bases (similar to BB84 which we have seen) as the qubits $H^{\theta_1} |x_1\rangle, \ldots, H^{\theta_{2n}} |x_{2n}\rangle$.
2. B measures each received qubit in a random basis (computational or Hadamard). The bases used by B are recorded as $\theta' = \theta'_1 \ldots \theta'_{2n}$. Outcomes are recorded in the string $x' = x'_1 \ldots x'_{2n}$.
3. Both A and B wait for time t (which waits long enough for B's quantum memory to "expire" (assuming it does!), so that B cannot have the received qubits before the next step – it must have measured the qubits and stored the outcomes by then – this is a security requirement since B needs to have measured its received qubits before knowing A's encoding basis string).
4. After time t, A assumes that B has done the measurements and sends to B the basis string θ and B determines the set of indices where it used the same basis as A, i.e. $S = \{j | \theta_j = \theta'_j\}$. B sets $S_y = S$ and $S_{1-y} = \{1, \ldots, 2n\} \backslash S$. B sends (S_0, S_1) to A – we assume that $|S_0| = |S_1| = n$, for large enough n.
5. A sends $d'_0 = d_0 \oplus x_{S_0}$ and $d'_1 = d_1 \oplus x_{S_1}$ to B, where $\oplus$ is the bitwise addition modulo 2, and $x_{S_0} = x_{k_1} \ldots x_{k_n}$, where $k_i \in S_0$, and x_{S_1} similarly defined. Note that d'_0 and d'_1 are encrypted values of d_0 and d_1 and B can only decrypt one of them.
6. A then outputs $\perp$ and B outputs $d'_y \oplus x'_{S_y}$.

We can see that the protocol correctly computes f_A and f_B above: A outputs $\perp$ at the end of the protocol, and B computes what it wanted:

$$d'_y \oplus x'_{S_y} = (d_y \oplus x_{S_y}) \oplus x'_{S_y} = d_y$$

since we know that $x_{S_y} \oplus x'_{S_y} = 0$, i.e. $x_{S_y} = x'_{S_y}$, because $S = \{j | \theta_j = \theta'_j\}$, and $S_y = S$ from step (4) above, the same basis was used by A and B for all those bits indexed in S_y. Vidick and Wehner provide a more detailed discussion of the security properties, but we can see that in step (3), if they did not wait long enough, B could hold the qubits in memory till it received the encoding bases θ from A and then just measure all the qubits in the same bases as A, then sends $(S_0 = \{1, \ldots, n\}, S_1 = \{n+1, \ldots, 2n\})$ to A like in step (4), and when A sends d'_0 and d'_1 to B like in step (5), B can compute both d_0 and d_1 since $x_{S_y} \oplus x'_{S_y} = 0$ for both $y = 0$ and $y = 1$! Another way for A to be sure that B has done the measurements before it sends it θ is for B to measure and commit its measured bits using the bit commitment

18 See the excellent Week8 and Week9 notes on OT by Wehner and Vidick at https://ocw .tudelft.nl/courses/quantum-cryptography. Their book on "Introduction to Quantum Cryptography" also has a deeper discussion on OT.

protocol. *A* does not know what *B* has asked for since in step (5) it sent both data items to *B*, though in encrypted form. Note that the protocol has similarities with the BB84 protocol since it basically uses the bits measured in the same bases as a shared key. There are also generalizations to 1-N OT, e.g. a review of quantum OT is [Santos et al., 2022].

3.2.9 Discussion

We have looked at a number of quantum protocols among multiple parties. Many of the protocols have a cryptographic nature and discussed extensively in quantum cryptography, but we outlined some of them as they would be useful in distributed computing settings. These are also rather "classical" protocols and there have been many new developments since. But it is hoped that this collection provides an overview of a range of quantum protocols and a taste for the highly technical reasoning involved.

3.3 Summary

In general, entanglement is not a replacement for communication. But we have seen that, if classical communication is disallowed, entanglement can enable certain distributed computations not possible without entanglement, and that, if classical communication is allowed, then, with entanglement, fewer classical bits of communication would be required to perform certain distributed computations. But the reduction in communication complexity is not always the case, i.e. not for any distributed computation.[19] We have also looked at many interesting quantum protocols, with analogs in the classical distributed computing setting. There are other more advanced protocols for the problems we have seen and some protocols which we did not consider, such as quantum multiparty computation. We saw that the sharing of entangled states is a key idea in many of these protocols. Deeper analysis of the protocols is also given in the original cited papers, which we did not provide in this chapter.

References

Dorit Aharonov, André Chailloux, Maor Ganz, Iordanis Kerenidis, and Loïck Magnin. A simpler proof of the existence of quantum weak coin flipping with arbitrarily small bias. *SIAM Journal on Computing*, 45(3):633–679, 2016.

---

19 For example, the work by Elkin et al. [2014] showed that the ability of nodes (in a network) to share entanglement and to send quantum information cannot help significantly to speed up (compared to the classical setting) distributed algorithms for certain problems such as computing the minimum spanning tree, and shortest paths.

Andris Ambainis. A new protocol and lower bounds for quantum coin flipping. *Journal of Computer and System Sciences*, 68(2):398–416, 2004.

A. Ambainis, H. Buhrman, Y. Dodis, and H. Rohrig. Multiparty quantum coin flipping. In *Proceedings of the 19th IEEE Annual Conference on Computational Complexity, 2004*, pages 250–259, 2004. doi: 10.1109/CCC.2004.1313848.

Myrto Arapinis, Nikolaos Lamprou, Elham Kashefi, and Anna Pappa. Definitions and security of quantum electronic voting. *ACM Transactions on Quantum Computing*, 2(1), Apr 2021. ISSN 2643-6809. doi: 10.1145/3450144.

Atul Singh Arora, Jérémie Roland, and Stephan Weis. Quantum weak coin flipping. In *Proceedings of the 51st Annual ACM SIGACT Symposium on Theory of Computing, STOC 2019*, pages 205–216, New York, NY, USA, 2019. Association for Computing Machinery. ISBN 9781450367059. doi: 10.1145/3313276.3316306.

Michael Ben-Or and Avinatan Hassidim. Fast quantum byzantine agreement. In *Proceedings of the Thirty-Seventh Annual ACM Symposium on Theory of Computing, STOC '05*, pages 481–485, New York, NY, USA, 2005. Association for Computing Machinery. ISBN 1581139608. doi: 10.1145/1060590.1060662.

Guido Berlín, Gilles Brassard, Félix Bussières, Nicolas Godbout, Joshua A. Slater, and Wolfgang Tittel. Experimental loss-tolerant quantum coin flipping. *Nature Communications*, 2(1):561, 2011.

Manuel Blum. Coin flipping by telephone a protocol for solving impossible problems. *SIGACT News*, 15(1):23–27, Jan 1983. ISSN 0163-5700. doi: 10.1145/1008908.1008911.

Anne Broadbent and Alain Tapp. Information-theoretic security without an honest majority. In Kaoru Kurosawa, editor, *Advances in Cryptology –ASIACRYPT 2007*, pages 410–426, Berlin, Heidelberg, 2007. Springer-Verlag.

Harry Buhrman, Richard Cleve, Serge Massar, and Ronald de Wolf. Nonlocality and communication complexity. *Reviews of Modern Physics*, 82:665–698, Mar 2010. doi: 10.1103/RevModPhys.82.665.

A. Chailloux and I. Kerenidis. Optimal quantum strong coin flipping. In *IEEE 50th Annual Symposium on Foundations of Computer Science (FOCS 2009)*, Los Alamitos, CA, USA, Oct 2009. IEEE Computer Society. doi: 10.1109/FOCS.2009.71.

David Chaum. The dining cryptographers problem: unconditional sender and recipient untraceability. *Journal of Cryptology*, 1(1):65–75, 1988.

Yi-Peng Chen, Jing-Yang Liu, Ming-Shuo Sun, Xing-Xu Zhou, Chun-Hui Zhang, Jian Li, and Qin Wang. Experimental measurement-device-independent quantum key distribution with the double-scanning method. *Optics Letters*, 46(15):3729–3732, Aug 2021.

Vicent Cholvi. Quantum byzantine agreement for any number of dishonest parties. *Quantum Information Processing*, 21(4):151, 2022.

Matthias Christandl and Stephanie Wehner. Quantum anonymous transmissions. In Bimal Roy, editor, *Advances in Cryptology - ASIACRYPT 2005*, pages 217–235, Berlin, Heidelberg, 2005. Springer-Verlag.

Richard Cleve and Harry Buhrman. Substituting quantum entanglement for communication. *Physical Review A*, 56:1201–1204, Aug 1997. doi: 10.1103/PhysRevA.56.1201.

Richard Cleve, Daniel Gottesman, and Hoi-Kwong Lo. How to share a quantum secret. *Physical Review Letters*, 83:648–651, Jul 1999. doi: 10.1103/PhysRevLett.83.648.

C. Döscher and M. Keyl. An introduction to quantum coin tossing. *Fluctuations and Noise Letters*, 02(04):R125–R137, 2002. Available at https://arxiv.org/abs/quant-ph/0206088.

Artur K. Ekert. Quantum cryptography based on Bell's theorem. *Physical Review Letters*, 67:661–663, Aug 1991. doi: 10.1103/PhysRevLett.67.661.

Michael Elkin, Hartmut Klauck, Danupon Nanongkai, and Gopal Pandurangan. Can quantum communication speed up distributed computation? In *Proceedings of the 2014 ACM Symposium on Principles of Distributed Computing, PODC '14*, pages 166–175, New York, NY, USA, 2014. Association for Computing Machinery. ISBN 9781450329446. doi: 10.1145/2611462.2611488.

Sascha Gaertner, Mohamed Bourennane, Christian Kurtsiefer, Adán Cabello, and Harald Weinfurter. Experimental demonstration of a quantum protocol for byzantine agreement and liar detection. *Physical Review Letters*, 100:070504, Feb 2008. doi: 10.1103/PhysRevLett.100.070504.

Maor Ganz. Quantum leader election. *Quantum Information Processing*, 16(3):73, 2017.

Nicolas Gisin, Grégoire Ribordy, Wolfgang Tittel, and Hugo Zbinden. Quantum cryptography. *Reviews of Modern Physics*, 74:145–195, Mar 2002. doi: 10.1103/RevModPhys.74.145.

Lov K. Grover. A fast quantum mechanical algorithm for database search. In *Proceedings of the Twenty-Eighth Annual ACM Symposium on Theory of Computing, STOC '96*, pages 212–219, New York, NY, USA, 1996. Association for Computing Machinery. ISBN 0897917855. doi: 10.1145/237814.237866.

Mark Hillery, Vladimír Bužek, and André Berthiaume. Quantum secret sharing. *Physical Review A*, 59:1829–1834, Mar 1999. doi: 10.1103/PhysRevA.59.1829.

Zixin Huang, Siddarth Koduru Joshi, Djeylan Aktas, Cosmo Lupo, Armanda O. Quintavalle, Natarajan Venkatachalam, Sören Wengerowsky, Martin Lončarić, Sebastian Philipp Neumann, Bo Liu, Željko Samec, Laurent Kling, Mario Stipčević, Rupert Ursin, and John G. Rarity. Experimental implementation of secure anonymous protocols on an eight-user quantum key distribution network. *npj Quantum Information*, 8(1):25, 2022.

S. Iblisdir and N. Gisin. Byzantine agreement with two quantum-key-distribution setups. *Physical Review A*, 70:034306, Sep 2004. doi: 10.1103/PhysRevA.70.034306.

I. Kerenidis and A. Nayak. Weak coin flipping with small bias. *Information Processing Letters*, 89(3):131–135, 2004.

Emil T. Khabiboulline, Juspreet Singh Sandhu, Marco Ugo Gambetta, Mikhail D. Lukin, and Johannes Borregaard. Efficient quantum voting with information-theoretic security. CoRR, abs/2112.14242, 2021. Available at https://arxiv.org/abs/2112.14242.

Leslie Lamport, Robert Shostak, and Marshall Pease. The byzantine generals problem. *ACM Transactions on Programming Languages and Systems*, 4(3):382–401, Jul 1982. ISSN 0164-0925. doi: 10.1145/357172.357176.

Yinxiang Long, Cai Zhang, and Zhiwei Sun. Standard (3, 5)-threshold quantum secret sharing by maximally entangled 6-qubit states. *Scientific Reports*, 11(1):22649, 2021.

Miralem Mehic, Stefan Rass, Peppino Fazio, and Miroslav Voznák. *Quantum Key Distribution Networks - A Quality of Service Perspective*, 1st edition. Springer, 2022. ISBN 978-3-031-06607-8. doi: 10.1007/978-3-031-06608-5.

C. Mochon. Quantum weak coin-flipping with bias of 0.192. In *45th Annual IEEE Symposium on Foundations of Computer Science*, pages 2–11, 2004. doi: 10.1109/FOCS.2004.55.

Carlos Mochon. Large family of quantum weak coin-flipping protocols. *Physical Review A*, 72:022341, Aug 2005. doi: 10.1103/PhysRevA.72.022341.

Carlos Mochon. Quantum weak coin flipping with arbitrarily small bias, 2007. Available at https://arxiv.org/abs/0711.4114.

Gláucia Murta, Federico Grasselli, Hermann Kampermann, and Dagmar Bruß. Quantum conference key agreement: a review. *Advanced Quantum Technologies*, 3(11):2000025, 2020.

Anna Pappa, Paul Jouguet, Thomas Lawson, André Chailloux, Matthieu Legré, Patrick Trinkler, Iordanis Kerenidis, and Eleni Diamanti. Experimental plug and play quantum coin flipping. *Nature Communications*, 5(1):3717, 2014.

Manuel B. Santos, Paulo Mateus, and Armando N. Pinto. Quantum oblivious transfer: a short review. *Entropy*, 24(7):945, 2022.

R. W. Spekkens and Terry Rudolph. Quantum protocol for cheat-sensitive weak coin flipping. *Physical Review Letters*, 89:227901, Nov 2002. doi: 10.1103/PhysRevLett.89.227901.

Seiichiro Tani, Hirotada Kobayashi, and Keiji Matsumoto. Exact quantum algorithms for the leader election problem. *ACM Transactions on Computation Theory*, 4(1), Mar 2012. ISSN 1942-3454. doi: 10.1145/2141938.2141939.

Chia-Wei Tsai, Chun-Wei Yang, and Jason Lin. Multiparty mediated quantum secret sharing protocol. *Quantum Information Processing*, 21(2):63, 2022.

Gilles Van Assche. *Quantum Cryptography and Secret-Key Distillation*. Cambridge University Press, 2006. ISBN 978-0-521-86485-5.

R. Wolf. *Quantum Key Distribution: An Introduction with Exercises*. Lecture Notes in Physics. Springer International Publishing, 2021. ISBN 9783030739911. URL https://books.google.com.au/books?id=sM47EAAAQBAJ.

Huanguo Zhang, Zhaoxu Ji, Houzhen Wang, and Wanqing Wu. Survey on quantum information security. *China Communications*, 16(10):1–36, 2019. doi: 10.23919/JCC.2019.10.001.

4

Distributed Quantum Computing – Distributed Control of Quantum Gates

This chapter looks at distributed quantum computing from a different perspective compared to Chapter 3. We will look at computations described by quantum circuits involving multiple qubits, where the qubits reside on multiple nodes, each node hosting one or more of the qubits. This chapter discusses nonlocal gates, or quantum gates for computations over several nodes. In particular, the chapter discusses the distributed CNOT gate, and different ways for its implementation, and discusses distributed control of gates beyond the CNOT gate. The discussion in this chapter is mainly based on the work by Yimsiriwattana and Lomonaco [2005] and earlier work by Eisert et al. [2000]. The CNOT gate is significant since using this gate and single qubit gates (or some suitable selection of gates as we saw in Chapter 2), one can achieve a universal gate set and realize a universal quantum computer. Hence, with single qubit gates (which would be local to the respective nodes on which the qubits reside) together with distributed CNOT operations, one effectively covers the range of possible distributed quantum circuit computations. The chapter also reviews briefly work on distributing quantum circuits and architectures for distributed control and management for distributed quantum computers.

4.1 Performing a Distributed CNOT

Consider a distributed CNOT operation where the control qubit is on one machine (say machine A) and the target qubit is on another machine (i.e. B). How can this be done?

4.1.1 Using Teleportation

Recall teleportation from Chapter 3 where a qubit state can be moved from one machine to another. So, one way is to teleport the qubit from A to B and then use

From Distributed Quantum Computing to Quantum Internet Computing: An Introduction,
First Edition. Seng W. Loke.
© 2024 The Institute of Electrical and Electronics Engineers, Inc. Published 2024 by John Wiley & Sons, Inc.

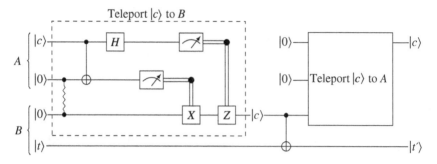

Figure 4.1 Distributed CNOT using teleportation.

the teleported qubit as control to affect the target qubit at B – the teleportation will involve one entangled pair of qubits shared between A and B, and two classical bits of information to be transferred from A to B as well. Finally, the control qubit can be teleported back from A to B so that A gets the control qubit back – we need this step since in a typical CNOT gate, the control qubit is unaffected and not lost.

We now look at a quantum circuit for this method – imagining that A has two qubits (the top two lines) and B has two qubits (the bottom two lines) and supposing that A and B are on two different machines. The circuit can be of the form shown in Figure 4.1, which first teleports the control qubit to B, performs the CNOT at B locally, and then teleporting the control qubit (back) to A. Here, we have a distributed CNOT operation performed, that is, if $|c\rangle = \alpha|0\rangle + \beta|1\rangle$, with $|\alpha|^2 + |\beta|^2 = 1$, we can write an expression for the distributed CNOT involving $|t\rangle_B$ as follows (adding subscripts to show location):

$$distCNOT(|c\rangle_A \otimes |t\rangle_B)$$
$$= distCNOT((\alpha|0\rangle_A + \beta|1\rangle_A) \otimes |t\rangle_B)$$
$$= \alpha|0\rangle_A|t\rangle_B + \beta|1\rangle_A(X|t\rangle_B)$$

Note that two teleportation operations are used, to teleport a qubit from A to B and then from B to A, so that we utilize two entangled pairs of qubits and four classical bits between A and B (i.e. one pair of entangled qubits and two classical bits for each teleportation).

4.1.2   A More Efficient Method With Cat-Like States

But is there a more efficient way? Yes, as shown by Eisert et al. [2000]. The circuit is as shown in Figure 4.2.

In fact, Eisert et al. also proved that this is optimal for nonlocal implementation of a quantum CNOT gate. Let's see how this works.

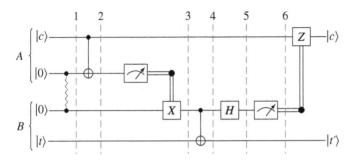

Figure 4.2 Distributed CNOT by Eisert et al.

Consider the state of the system at slice 1 above, where we let $|c\rangle = \alpha\,|0\rangle + \beta\,|1\rangle$:

$$\text{at slice 1}: \quad (\alpha\,|0\rangle + \beta\,|1\rangle) \otimes \left(\frac{1}{\sqrt{2}}\,|00\rangle + \frac{1}{\sqrt{2}}\,|11\rangle \right) \otimes |t\rangle$$

$$= \left(\frac{\alpha}{\sqrt{2}}\,|000\rangle + \frac{\alpha}{\sqrt{2}}\,|011\rangle + \frac{\beta}{\sqrt{2}}\,|100\rangle + \frac{\beta}{\sqrt{2}}\,|111\rangle \right) \otimes |t\rangle$$

then, applying the CNOT between the top two qubits, we have, at slice 2:

$$\text{at slice 2}: \quad \left(\frac{\alpha}{\sqrt{2}}\,|000\rangle + \frac{\alpha}{\sqrt{2}}\,|011\rangle + \frac{\beta}{\sqrt{2}}\,|110\rangle + \frac{\beta}{\sqrt{2}}\,|101\rangle \right) \otimes |t\rangle$$

Then, the second qubit (from the top) is measured, and

1) if we get a 0, and this is sent from A to B, and since it is 0, the X gate is not applied, then, at slice 3, we have:

$$\text{at slice 3}: \quad \sqrt{2} \cdot \left(\frac{\alpha}{\sqrt{2}}\,|000\rangle + \frac{\beta}{\sqrt{2}}\,|101\rangle \right) \otimes |t\rangle$$

$$= (\alpha\,|000\rangle + \beta\,|101\rangle) \otimes |t\rangle$$

since $\dfrac{1}{\sqrt{prob(0)\ in\ second\ qubit}} = \dfrac{1}{\sqrt{\left|\frac{\alpha}{\sqrt{2}}\right|^2 + \left|\frac{\beta}{\sqrt{2}}\right|^2}} = \dfrac{1}{\sqrt{1/2}} = \sqrt{2}$, and then applying the

CNOT between the two qubits at B locally:

$$\text{at slice 4}: \quad \alpha\,|000\rangle\,|t\rangle + \beta\,|101\rangle\,(X\,|t\rangle)$$

and then applying gate H in the third line (from the top):

$$\text{at slice 5}: \quad \alpha\,|00\rangle\,(H\,|0\rangle)\,|t\rangle + \beta\,|10\rangle\,(H\,|1\rangle)(X\,|t\rangle)$$

$$= \frac{\alpha}{\sqrt{2}}\,|000\rangle\,|t\rangle + \frac{\alpha}{\sqrt{2}}\,|001\rangle\,|t\rangle + \frac{\beta}{\sqrt{2}}\,|100\rangle\,(X\,|t\rangle)$$

$$- \frac{\beta}{\sqrt{2}}\,|101\rangle\,(X\,|t\rangle)$$

Then, measuring the qubit at B (at third line from the top):

a) suppose we get 0, we have:

$$\text{at slice 6}: \quad \alpha\,|000\rangle\,|t\rangle + \beta\,|100\rangle\,(X\,|t\rangle)$$

$$= \alpha\,|0\rangle\,|00\rangle\,|t\rangle + \beta\,|1\rangle\,|00\rangle\,(X\,|t\rangle)$$

and "0" sent to A means that the Z gate is not applied, and so, leaving out the second and third qubits, and adding subscripts, we can write the final state as

$$\alpha|0\rangle_A|t\rangle_B + \beta|1\rangle_A(X|t\rangle_B)$$

b) suppose we get 1, we then have:

$$\text{at slice 6}: \quad \alpha\,|001\rangle\,|t\rangle - \beta\,|101\rangle\,(X\,|t\rangle)$$

$$= \alpha\,|0\rangle\,|01\rangle\,|t\rangle - \beta\,|1\rangle\,|01\rangle\,(X\,|t\rangle)$$

and "1" sent to A means that the Z gate is applied, and so, leaving out the second and third qubits, and adding subscripts, we can write the final state as

$$\alpha|0\rangle_A|t\rangle_B + \beta|1\rangle_A(X|t\rangle_B)$$

which is the same as in case (a).

2) and if we get a 1, and this is sent from A to B, and since it is 1, the X gate is applied, then, at slice 3, we have:

$$\text{at slice 3}: \quad (\alpha\,|010\rangle + \beta\,|111\rangle) \otimes |t\rangle$$

and then applying the CNOT between the two qubits at B locally:

$$\text{at slice 4}: \quad \alpha\,|010\rangle\,|t\rangle + \beta\,|111\rangle\,(X\,|t\rangle)$$

and then applying gate H in the third line (from the top):

$$\text{at slice 5}: \quad \alpha\,|01\rangle\,(H\,|0\rangle)\,|t\rangle + \beta\,|11\rangle\,(H\,|1\rangle)(X\,|t\rangle)$$

$$= \frac{\alpha}{\sqrt{2}}\,|010\rangle\,|t\rangle + \frac{\alpha}{\sqrt{2}}\,|011\rangle\,|t\rangle + \frac{\beta}{\sqrt{2}}\,|110\rangle\,(X\,|t\rangle)$$

$$- \frac{\beta}{\sqrt{2}}\,|111\rangle\,(X\,|t\rangle)$$

Then, measuring the qubit at B (the third qubit or third line from the top):

a) suppose we get 0, we have:

$$\text{at slice 6}: \quad \alpha\,|010\rangle\,|t\rangle + \beta\,|110\rangle\,(X\,|t\rangle)$$

$$= \alpha\,|0\rangle\,|10\rangle\,|t\rangle + \beta\,|1\rangle\,|10\rangle\,(X\,|t\rangle)$$

and "0" sent to A means that the Z gate is not applied, and so, leaving out the second and third qubits, and adding subscripts, we can write the final state as

$$\alpha|0\rangle_A|t\rangle_B + \beta|1\rangle_A(X|t\rangle_B)$$

b) suppose we get 1, we then have:

$$\text{at slice 6}: \quad \alpha\,|011\rangle\,|t\rangle - \beta\,|111\rangle\,(X\,|t\rangle)$$

$$= \alpha\,|0\rangle\,|11\rangle\,|t\rangle - \beta\,|1\rangle\,|11\rangle\,(X\,|t\rangle)$$

and "1" sent to A means that the Z gate is applied, and so, leaving out the second and third qubits, and adding subscripts, we can write the final state as

$$\alpha|0\rangle_A|t\rangle_B + \beta|1\rangle_A(X|t\rangle_B)$$

which is the same as in case (a).

Note that the cases (1)(a), (1)(b), (2)(b), and (2)(a) all have the same result if we consider just the first and fourth qubits, which is also the same as the result from the teleportation method, which is what we expect – we now have a more efficient way to implement a distributed CNOT. Note that in this method, the distributed CNOT operation was done using one pair of entangled qubits (i.e. one *ebit*) shared between A and B and two classical bits (one from A to B and the other from B to A), which is half the resources required for the teleportation method! It turns out that this is optimal as shown by Eisert et al. [2000], in that one bit of classical communication from A to B, one classical bit from B to A and one shared ebit are together what is minimal to implement a distributed CNOT gate.

Let us take a closer look at why the above has worked. One main observation is that the first and third qubits (as underlined below) form a *cat-like state* at slice 3 in both cases:

$$\text{case (1)}: \quad (\alpha\,|\underline{0}0\underline{0}\rangle + \beta\,|\underline{1}0\underline{1}\rangle) \otimes |t\rangle$$

$$\text{case (2)}: \quad (\alpha\,|\underline{0}1\underline{0}\rangle + \beta\,|\underline{1}1\underline{1}\rangle) \otimes |t\rangle$$

that is, leaving out the second qubit, the cat-like state in both cases is

$$\alpha\,|00\rangle\,|t\rangle + \beta\,|11\rangle\,(X\,|t\rangle)$$

from the underlined qubits, where we see that the first and third qubits always have the same value (0 or 1) – which means that controlling on the first qubit (at A) is the same as controlling on the third qubit (i.e. the first qubit at B). Now, adding the subscripts to denote the location of the first and third qubits, and the target qubit, in slice 3:

$$\alpha|0\rangle_A|0\rangle_B|t\rangle_B + \beta|1\rangle_A|1\rangle_B(X|t\rangle_B)$$

We see that step 4 (the local CNOT between the third and fourth qubits at B) is effectively a CNOT between the first qubit (at A) and the fourth qubit (at B)!

Distribution of such cat-like states is key to being able to implement distributed control gates using entanglement as a resource. In general, we can summarize this concept via the following circuit – where the second qubit is omitted.

$$|c\rangle = \alpha\,|00\rangle + \beta\,|11\rangle\,\left\{\vphantom{\Big|}\right.$$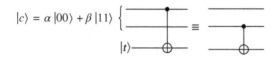

If the control qubit was any one of the two qubits in the cat-like state $\alpha\,|00\rangle + \beta\,|11\rangle$, the target will be affected the same way, since these two qubits will have effectively the same state, both are 0 with probability $|\alpha|^2$ or both are 1 with probability $|\beta|^2$. This means that if the cat-like state $|c\rangle$ has its two qubits distributed on different machines, then one can control on either of them. Note that the purpose of applying the H gate (i.e. the steps from slice 4 onward in Figure 4.2) and measuring and sending the measurement to control on Z is to disentangle the first qubit (at A) from the third qubit (i.e. the first qubit of B), now that the distributed CNOT has been done.

4.2 Beyond the Distributed CNOT

We can generalize the above in a number of ways:

- using the same control qubit for multiple target qubits on different machines;
- using multiple control qubits on different machines for the same target qubit; and
- going beyond performing the NOT on the target qubit to performing a general unitary operation U.

which we will examine in this section.

4.2.1 Same Control Qubit for Multiple Target Qubits on Different Machines

Suppose we want to use the control qubit in A on target qubits in more than one other machine. Certainly, one way to do this is to teleport the control qubit to each machine in turn, and on each machine, control on the received qubit. However, a more efficient way uses the generalizations of such cat-like states to more than two qubits, taking the form of GHZ states mentioned in Chapter 2. Consider the example with a three qubit cat-like state as follows.

$$|c\rangle = \alpha\,|000\rangle + \beta\,|111\rangle\,\left\{\vphantom{\Big|}\right.$$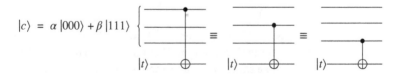

If the control qubit was any one of the three qubits in the cat-like state $\alpha\,|000\rangle + \beta\,|111\rangle$, then the target will be affected the same way, since these three qubits will have the same state, they are concurrently 0 with probability $|\alpha|^2$ or are 1 with probability $|\beta|^2$.

Let's see how this works in detail. Extending from the single target case, let's consider another machine C that also has a qubit $|s\rangle$ to be controlled by the qubit $|c\rangle$ at A, as shown in Figure 4.3. Note that the three nodes share a (distributed) GHZ state (one qubit each) marked by the wavy lines across the three nodes.[1]

Now consider the state of the system at slice 1 in Figure 4.3, where we let $|c\rangle = \alpha|0\rangle + \beta|1\rangle$, and reordering the qubits in our expressions, i.e. writing the target qubit of B just before the target qubit of C:

$$\text{at slice 1}: \quad (\alpha|0\rangle + \beta|1\rangle) \otimes \left(\frac{1}{\sqrt{2}}|000\rangle + \frac{1}{\sqrt{2}}|111\rangle \right) \otimes |t\rangle \otimes |s\rangle$$

$$= \left(\frac{\alpha}{\sqrt{2}}|0000\rangle + \frac{\alpha}{\sqrt{2}}|0111\rangle + \frac{\beta}{\sqrt{2}}|1000\rangle + \frac{\beta}{\sqrt{2}}|1111\rangle \right)$$

$$\otimes |t\rangle \otimes |s\rangle$$

then, applying the CNOT between the top two qubits, we have, at slice 2:

$$\text{at slice 2}: \quad \left(\frac{\alpha}{\sqrt{2}}|0000\rangle + \frac{\alpha}{\sqrt{2}}|0111\rangle + \frac{\beta}{\sqrt{2}}|1100\rangle + \frac{\beta}{\sqrt{2}}|1011\rangle \right)$$

$$\otimes |t\rangle \otimes |s\rangle$$

---

1 A distributed *GHZ* state can be created by fusing multiple Bell pairs, as mentioned in Chapter 2. For example, if we have the (distributed) Bell pairs between A and B (qubits a and b_1) and between B and C (qubits b_2 and c) as follows: $(|00\rangle_{ab_1} + |11\rangle_{ab_1})/\sqrt{2}$ and $(|00\rangle_{b_2c} + |11\rangle_{b_2c})/\sqrt{2}$, then one can carry out the following operations at B and C (Pauli gate correction X^m depending on the measurement outcome $m \in \{0, 1\}$ of b_2, sent from B to C) to merge the two Bell pairs to form a *GHZ* state:

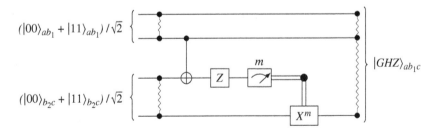

The resulting state is a *GHZ* state with qubits distributed across A, B, and C. Protocols to create n-qubit *GHZ* states is discussed in detail in de Bone et al. [2020]. Such a circuit (which, including the components to generate the two Bell pairs from initial states (two pairs of $|00\rangle$), can be encapsulated as a gate) might be called an *entangling gate*, denoted by E_3, or more generally, E_n for generating $|GHZ_n\rangle$. Note that the Bell pairs would each be distributed across two nodes.

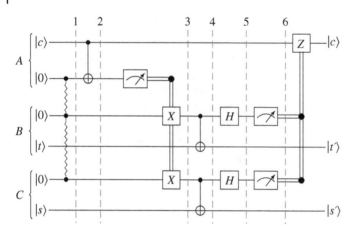

Figure 4.3 Distributed CNOT with single control qubit (on *A*) for multiple target qubits (one on *B* and one on *C*).

Then, the second qubit (from the top) is measured, and

1) if we get a 0, and this is sent from *A* to *B* and *C*, and since it is 0, the *X* gate is not applied at both *B* and *C*, then, at slice 3, we have:

$$\text{at slice 3}: \quad \sqrt{2} \cdot \left(\frac{\alpha}{\sqrt{2}} |0000\rangle + \frac{\beta}{\sqrt{2}} |1011\rangle \right) \otimes |t\rangle \otimes |s\rangle$$

$$= (\alpha |0000\rangle + \beta |1011\rangle) \otimes |t\rangle \otimes |s\rangle$$

since $\dfrac{1}{\sqrt{prob(0) \text{ in second qubit}}} = \dfrac{1}{\sqrt{|\frac{\alpha}{\sqrt{2}}|^2 + |\frac{\beta}{\sqrt{2}}|^2}} = \dfrac{1}{\sqrt{1/2}} = \sqrt{2}$, and then applying the CNOT between the two qubits at *B* and at *C*, both locally:

$$\text{at slice 4}: \quad \alpha |0000\rangle |t\rangle |s\rangle + \beta |1011\rangle (X |t\rangle)(X |s\rangle)$$

and then applying gate *H* in the third line and fifth line (from the top):

$$\text{at slice 5}: \quad \alpha |00\rangle (H |0\rangle)(H |0\rangle) |t\rangle |s\rangle + \beta |10\rangle (H |1\rangle)(H |1\rangle)(X |t\rangle)(X |s\rangle)$$

$$= \frac{\alpha}{2} |0000\rangle |t\rangle |s\rangle + \frac{\alpha}{2} |0001\rangle |t\rangle |s\rangle$$

$$+ \frac{\alpha}{2} |0010\rangle |t\rangle |s\rangle + \frac{\alpha}{2} |0011\rangle |t\rangle |s\rangle$$

$$+ \frac{\beta}{2} |1000\rangle (X |t\rangle)(X |s\rangle) - \frac{\beta}{2} |1001\rangle (X |t\rangle)(X |s\rangle)$$

$$- \frac{\beta}{2} |1010\rangle (X |t\rangle)(X |s\rangle) + \frac{\beta}{2} |1011\rangle (X |t\rangle)(X |s\rangle)$$

Then, measuring the qubits at B and C (at third and fifth line from the top), suppose we get 0 at both B and C, we have:

at slice 6 : $\alpha\,|0000\rangle\,|t\rangle\,|s\rangle + \beta\,|1000\rangle\,(X\,|t\rangle)(X\,|s\rangle)$

$$= \alpha\,|0\rangle\,|000\rangle\,|t\rangle\,|s\rangle + \beta\,|1\rangle\,|000\rangle\,(X\,|t\rangle)(X\,|s\rangle)$$

and both "0"s sent to A which XORs (or sum modulo 2) the results to get 0 meaning that the Z gate is not applied, and so, leaving out the second, third and fourth qubits, and adding subscripts, we can write the final state as

$$\alpha|0\rangle_A|t\rangle_B|s\rangle_C + \beta|1\rangle_A(X|t\rangle_B)(X|s\rangle_C)$$

where we observe that the control qubit at A has been similarly applied to the target qubits at both B and C. A similar reasoning can be applied for other combinations of results of measurements at B and C, and then at slice 6, we get states as summarized in the table below.

Measured at B	Measured at C	At slice 6
0	0	$\alpha\,|0\rangle\,|000\rangle\,|t\rangle\,|s\rangle + \beta\,|1\rangle\,|000\rangle\,(X\,|t\rangle)(X\,|s\rangle)$
0	1	$\alpha\,|0\rangle\,|001\rangle\,|t\rangle\,|s\rangle - \beta\,|1\rangle\,|001\rangle\,(X\,|t\rangle)(X\,|s\rangle)$
1	0	$\alpha\,|0\rangle\,|010\rangle\,|t\rangle\,|s\rangle - \beta\,|1\rangle\,|010\rangle\,(X\,|t\rangle)(X\,|s\rangle)$
1	1	$\alpha\,|0\rangle\,|011\rangle\,|t\rangle\,|s\rangle + \beta\,|1\rangle\,|011\rangle\,(X\,|t\rangle)(X\,|s\rangle)$

Note that in the cases where the measurements from B and C are "0" and "1," and "1" and "0," respectively, then at A, the sum (modulo 2) is 1 so that the Z gate in the first line is applied. That is, e.g. in the case of the results being "0" at B and "1" at C, the final state, after applying the Z gate (which effectively converts the "$-$" in front of the β to "$+$") is $\alpha\,|0\rangle\,|001\rangle\,|t\rangle\,|s\rangle + \beta\,|1\rangle\,|001\rangle\,(X\,|t\rangle)(X\,|s\rangle)$, and so, leaving out the second, third and fourth qubits, and adding subscripts, we get the final state: $\alpha|0\rangle_A|t\rangle_B|s\rangle_C + \beta|1\rangle_A(X|t\rangle_B)(X|s\rangle_C)$, which is the same result as when the measurement results are "0" at both B and C.

2) and if we get a 1, and this is sent from A to B and C, and since it is 1, the X gate is applied at both B and C, then, at slice 3, we have:

at slice 3 : $(\alpha\,|0100\rangle + \beta\,|1111\rangle) \otimes |t\rangle \otimes |s\rangle$

and then applying the CNOT between the two qubits at B and C, locally:

at slice 4 : $\alpha\,|0100\rangle\,|t\rangle\,|s\rangle + \beta\,|1111\rangle\,(X\,|t\rangle)(X\,|s\rangle)$

and then applying gate H in the third line and fifth line (from the top):

$$at\ slice\ 5: \quad \alpha\,|01\rangle\,(H\,|0\rangle)(H\,|0\rangle)\,|t\rangle\,|s\rangle + \beta\,|11\rangle\,(H\,|1\rangle)(H\,|1\rangle)(X\,|t\rangle)(X\,|s\rangle)$$

$$= \frac{\alpha}{2}\,|0100\rangle\,|t\rangle\,|s\rangle + \frac{\alpha}{2}\,|0101\rangle\,|t\rangle\,|s\rangle$$

$$+ \frac{\alpha}{2}\,|0110\rangle\,|t\rangle\,|s\rangle + \frac{\alpha}{2}\,|0111\rangle\,|t\rangle\,|s\rangle$$

$$+ \frac{\beta}{2}\,|1100\rangle\,(X\,|t\rangle)(X\,|s\rangle) - \frac{\beta}{2}\,|1101\rangle\,(X\,|t\rangle)(X\,|s\rangle)$$

$$- \frac{\beta}{2}\,|1110\rangle\,(X\,|t\rangle)(X\,|s\rangle) + \frac{\beta}{2}\,|1111\rangle\,(X\,|t\rangle)(X\,|s\rangle)$$

Then, measuring the qubits at B and C (at third and fifth line from the top), and depending on the outcomes, we can get the states as summarized in the following table.

Measured at B	Measured at C	At slice 6
0	0	$\alpha\,|0\rangle\,|100\rangle\,|t\rangle\,|s\rangle + \beta\,|1\rangle\,|100\rangle\,(X\,|t\rangle)(X\,|s\rangle)$
0	1	$\alpha\,|0\rangle\,|101\rangle\,|t\rangle\,|s\rangle - \beta\,|1\rangle\,|101\rangle\,(X\,|t\rangle)(X\,|s\rangle)$
1	0	$\alpha\,|0\rangle\,|110\rangle\,|t\rangle\,|s\rangle - \beta\,|1\rangle\,|110\rangle\,(X\,|t\rangle)(X\,|s\rangle)$
1	1	$\alpha\,|0\rangle\,|111\rangle\,|t\rangle\,|s\rangle + \beta\,|1\rangle\,|111\rangle\,(X\,|t\rangle)(X\,|s\rangle)$

and then applying the Z gate when required, and leaving out the second, third, and fourth qubits, and adding subscripts for location of qubits, we get the final state:

$$\alpha|0\rangle_A|t\rangle_B|s\rangle_C + \beta|1\rangle_A(X|t\rangle_B)(X|s\rangle_C)$$

which is the same result as case (1) above.

We point out the three qubit cat-like state at slice 3 in both cases (1) and (2) above:

$$case\ (1): \quad (\alpha\,|\underline{0000}\rangle + \beta\,|\underline{1011}\rangle) \otimes |ts\rangle$$

$$case\ (2): \quad (\alpha\,|\underline{0100}\rangle + \beta\,|\underline{1111}\rangle) \otimes |ts\rangle$$

leaving out the second qubit, we have, in both cases, the state: $(\alpha\,|000\rangle + \beta\,|111\rangle)\,|ts\rangle$, from the underlined qubits.

One can generalize further so that the control qubit is applied to target qubits on more than two machines as follows, e.g. target qubits on three machines: B, C, and D, we have the circuit shown in Figure 4.4 with the state of the form: $(\alpha\,|0000\rangle + \beta\,|1111\rangle)\,|tsr\rangle$ at slice 3 having the four qubit cat-like state, leaving out the second qubit. One can draw a similar circuit for n target qubits $|t_i\rangle$ on n machines, which uses the $(n+1)$ qubit cat-like state: $(\alpha|0\rangle^{\otimes(n+1)} + \beta|1\rangle^{\otimes(n+1)})\,|t_1\rangle \otimes \cdots \otimes |t_n\rangle$.

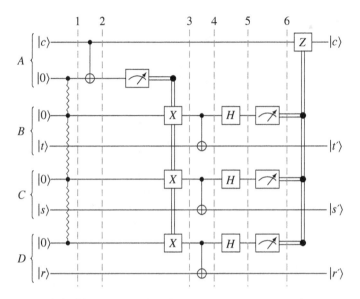

Figure 4.4 Distributed CNOT with single control qubit (on A) and three target qubits (on A, B, and C).

4.2.2 Multiple Control Qubits for the Same Target Qubit on a Different Machine

What if we wanted to have multiple controls (with each control qubit on a different machine) on one target qubit (on a machine different from those of the control qubits)? For example, we have an operation of the following form, where there are two control qubits and all needs to be "1" before the target qubit is flipped; otherwise, the target qubit is unchanged:

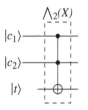

This is also called the CCNOT gate or the Toffoli gate. So, if we use the notation $\wedge_1(X)$ to represent the CNOT gate, then the above gate is denoted as $\wedge_2(X)$.

One way to implement this is to use teleportation. The control qubits are teleported from where they are (say, on different machines) to the machine containing the target qubit, and then the above operation is performed. And then the control qubits are teleported back to their origin machines. For two control qubits on different machines (from the target qubit), resources required for this would be two ebits and four classical bits for the two teleportation operations, and then another two ebits and four classical bits for teleporting the control qubits back to their origin machines, i.e. four ebits and eight classical bits.

A more efficient way is as follows also due to Eisert et al. [2000], as shown in Figure 4.5 for two control qubits, which uses four control classical bits and two ebits.

Assume that the control qubit at A is $|c1\rangle = \alpha_1|0\rangle_{A1} + \beta_1|1\rangle_{A1}$ and at B it is $|c2\rangle = \alpha_2|0\rangle_{B1} + \beta_2|1\rangle_{B1}$. Then, considering the qubits $A1$, $A2$, and $C1$, we start with $(\alpha_1|0\rangle_{A1} + \beta_1|1\rangle_{A1})|0\rangle_{A2}|0\rangle_{C1}$. Then, $A2$ and $C1$ are entangled, i.e. we have at slice 1: $(\alpha_1|0\rangle_{A1} + \beta_1|1\rangle_{A1}) \otimes (|00\rangle_{A2,C1} + |11\rangle_{A2,C1})/\sqrt{2}$. Then, at slice 2 (after the CNOT between $A1$ and $A2$), we have: $(\alpha_1|000\rangle + \alpha_1|011\rangle + \beta_1|110\rangle + \beta_1|101\rangle))/\sqrt{2}$. At slice 3, after measuring the second qubit $A2$ at A (second line):

1) If we get a 0, and this is sent from A to C, and since it is 0, the X gate on line 5 is not applied, then, at slice 3, we have:

$$at\ slice\ 3: \quad \alpha_1|0\rangle_{A1}|0\rangle_{A2}|0\rangle_{C1} + \beta_1|1\rangle_{A1}|0\rangle_{A2}|1\rangle_{C1}$$

with an expression showing only the qubits on A and the first qubit at C (i.e. $C1$) – the other qubits are unchanged at this point from the starting point.

2) If a 1 is obtained instead, and this is sent from A to C, and since it is 1, the X gate on line 5 is applied to qubit $C1$, then, at slice 3, we have:

$$at\ slice\ 3: \quad \alpha_1|0\rangle_{A1}|1\rangle_{A2}|0\rangle_{C1} + \beta_1|1\rangle_{A1}|1\rangle_{A2}|1\rangle_{C1}$$

and the other qubits are unchanged.

Note that if we omit writing $A2$ in both cases above, we get

$$at\ slice\ 3: \quad \alpha_1|0\rangle_{A1}|0\rangle_{C1} + \beta_1|1\rangle_{A1}|1\rangle_{C1}$$

that is, $A1$ has been effectively entangled with $C1$.

Then at, slice 6, after measuring the second qubit at B (line 4):

1) If we get a 0, and this is sent from B to C, and since it is 0, the X gate on line 6 is not applied, then, at slice 6, we have:

$$at\ slice\ 6: \quad \alpha_2|0\rangle_{B1}|0\rangle_{B2}|0\rangle_{C2} + \beta_2|1\rangle_{B1}|0\rangle_{B2}|1\rangle_{C2}$$

with an expression showing only the qubits on B and the second qubit at C (i.e. $C2$), while the qubits at A and $C1$ at slice 6 are as above when at slice 3 (since qubits at A and $C1$ are not changed between slice 3 and slice 6).

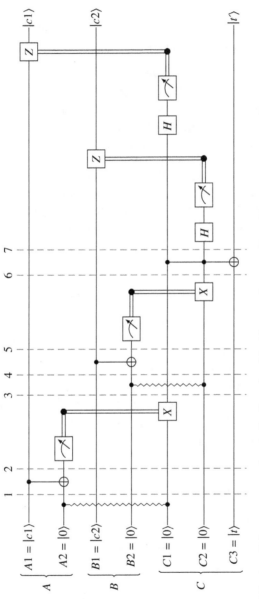

Figure 4.5 An efficient way to perform a distributed multiqubit control on a target qubit.

2) If a 1 is obtained instead, and this is sent from B to C, and since it is 1, the X gate on line 6 is applied, then, at slice 6, we have:

$$\text{at slice 6}: \quad \alpha_2|0\rangle_{B1}|1\rangle_{B2}|0\rangle_{C2} + \beta_2|1\rangle_{B1}|1\rangle_{B2}|1\rangle_{C2}$$

while the qubits at A and $C1$ at slice 6, are as above at slice 3.

Note that if we omit writing $B2$ in both cases above, we get

$$\text{at slice 6}: \quad \alpha_1|0\rangle_{B1}|0\rangle_{C2} + \beta_1|1\rangle_{B1}|1\rangle_{C2}$$

that is, $B1$ has been effectively entangled with $C2$.

Note that we want to have the target qubit $C3$ controlled by qubits $A1$ and $B1$. Due the entanglements of $A1$ with $C1$ and $B1$ with $C2$, this is equivalent to having the target qubit $C3$ controlling on the qubits on C ($C1$ and $C2$), at slice 6. This achieves the distributed CCNOT with control qubits $A1$ and $B1$ on target $C3$, via the local CCNOT with control qubits $C1$ and $C2$ on target $C3$,

Now that the local CCNOT has been done, to finish off, we should disentangle $A1$ from $C1$ and $B1$ from $C2$, as done via the H gates applied on $C1$ and $C2$ and the subsequent measurements sent to A and B to control on the respective Z gates. Note that we needed to "disentangle" in order that the control qubits are now decoupled from the target qubit.

4.2.3 Cat-Entangler and Cat-Disentangler Modules

Before going on to other operations beyond the X gate on the target, we observe several "modules" used in our previous circuits, which has been called the *cat-entangler* and the *cat-disentangler*.

A cat-entangler is a (sub-)circuit of the following form, which takes (i) a qubit $\alpha|0\rangle + \beta|1\rangle$ (on the top line) and (ii) a generalized GHZ state of the form $\frac{1}{\sqrt{2}}(|0\rangle^{\otimes n} + |1\rangle^{\otimes n})$ (which is represented by the other lines and is generated at slice 1 by the entangling gate E_n) and outputs a cat-like state $\alpha|0\rangle^{\otimes n} + \beta|1\rangle^{\otimes n}$ (when we leave out the second qubit).

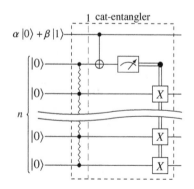

To see this, note that the output of the circuit is $(n + 1)$ qubits with a state of the form:

$$\alpha \,|0\rangle \otimes |x\rangle \otimes |0\rangle^{\otimes(n-1)} + \beta \,|1\rangle \otimes |x\rangle \otimes |1\rangle^{\otimes(n-1)}$$

(where x is either "0" or "1" depending on the measurement outcome on the second line), and leaving out the second qubit $|x\rangle$, we have the cat-like state:

$$\alpha \,|0\rangle \,|0\rangle^{\otimes(n-1)} + \beta \,|1\rangle \,|1\rangle^{\otimes(n-1)} = \alpha|0\rangle^{\otimes n} + \beta|1\rangle^{\otimes n}$$

And a cat-disentangler is of the form:

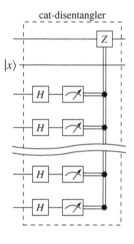

where the measurement outcomes from the lines are summed (mod 2) and the sum used as control to determine if the Z gate will be applied. The circuit will take a $(n + 1)$ qubit state: $\alpha \,|0\rangle \otimes |x\rangle \otimes |0\rangle^{\otimes(n-1)} + \beta \,|1\rangle \otimes |x\rangle \otimes |1\rangle^{\otimes(n-1)}$ (only the position of $|x\rangle$ indicated above) and output: $(\alpha \,|0\rangle + \beta \,|1\rangle) \otimes |x\rangle \otimes |s\rangle$, where s is an $(n - 1)$ bit string, i.e. $|s\rangle$ is an $(n - 1)$ qubit state due to the measurements from line 3 onwards – the effect is to disentangle the top qubit from the other qubits.

To illustrate these modules being used in circuits, we highlight in Figure 4.6 the cat-entangler and cat-disentangler modules as they were used in the circuits we have seen (i.e. Figures 4.2–4.5).

Note that we could replace the X gate in the circuits we have seen with some other operation, say U, and so, generalize to a distributed controlled-U operation.

4.3 Distributing Quantum Circuits and Compilation for Distributed Quantum Programs

Based on the ideas we have seen, a number of distributed versions of quantum algorithms have been developed, such as a distributed Shor algorithm

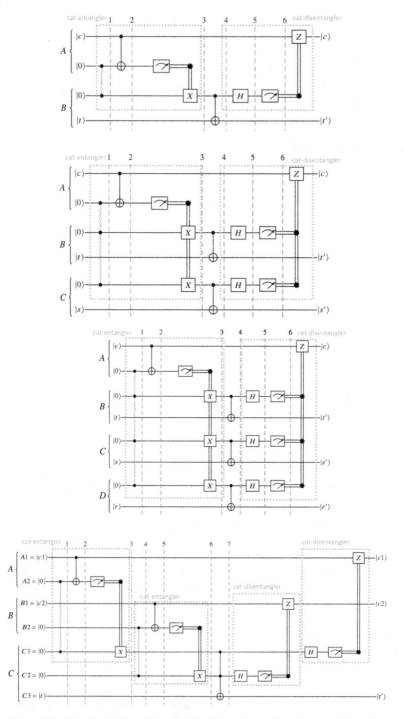

Figure 4.6 The circuits from Figures 4.2–4.5 marked with cat-entanglers and cat-disentanglers.

[Yimsiriwattana and Lomonaco Jr., 2004], a distributed algorithm for Simon's problem [Tan et al., 2022], and a distributed machine learning algorithm [Neumann and Wezeman, 2022] (see also [Kwak et al., 2022] for a review on quantum distributed deep learning).[2]

There has been work on taking a large quantum circuit and figuring out ways to partition the circuit or to distribute the circuit over a collection of nodes, with each node hosting one or more of the qubits, and their associated operations, using entanglement to link qubits on different nodes. Several considerations are required when distributing a quantum circuit and assigning qubits to nodes, e.g. designing (or finding alternative) quantum circuits to minimize internode teleportations (while still computing what is required) and since entanglement is a resource, one might try to keep entangled qubits between different nodes to a minimum when partitioning a circuit. We will not go into the details here but the reader can refer to Tang et al. [2021], Sundaram et al. [2021], Andrés-Martínez and Heunen [2019], Davarzani et al. [2022], and van Houte et al. [2020]. The reader can also look into [Cuomo et al., 2023] for a study on compiling quantum circuits for diferent topologies of nodes.

The idea of automating the distribution of a (typically large) quantum circuit into parts (a collection of smaller quantum circuits) which map to a given topology of quantum processing units (or quantum computers) brings us to the notion of *compilers* for distribute quantum programs. The compiler not only could optimize for reducing circuit depth but also reducing internode communications or the number of internode entangled pairs required (typically a relatively time-consuming process) while still performing the required computations, and such compilation should be done efficiently [Ferrari et al., 2021] – we come back to this point in Chapter 6.

4.4 Control and Scheduling for Distributed Quantum Computers

We saw in Chapter 1 the illustration of a distributed quantum computer in Figure 1.1, which showed how multiple nodes utilize both entanglement and classical communications for their operations.

Suppose that a large quantum circuit to be executed has been partitioned and generated is a resulting equivalent set of quantum circuits, each assigned to a node (with the ability to host and operate on one or more qubits as needed for its

---

2 This Website https://github.com/brunorijsman/quantum-internet-hackathon-2022 [last accessed: 8/3/2023] has Python-based implementations of the distributed quantum Fourier transformation on simulation platforms.

assigned circuit). As a small example, consider the example of the circuit shown in Figure 4.2 for a distributed CNOT where there are two circuits, one to be executed by node A and the other by B, and in terms of interaction and communication, requiring the resource of one shared EPR pair and sending of classical bits at the two steps, according to the diagram. How will this be executed?

Figure 4.7 illustrates (in a rather naive and high level way!) the execution of the four qubit circuit for a distributed CNOT. First, after deciding on the set of quantum circuits to execute, the respective quantum circuits together with classical control information (or a classical program) will need to be loaded at nodes A and B (separate programs on A and on B)[3] and executed with a mix of local operations and synchronization where there is required entanglement or transfer of bits. Note that an entangled pair between nodes A and B is required to be acquired from the quantum Internet as shown in the figure. This would require some classical communication and synchronization between the nodes so that subsequent local operations with the entangled qubits (e.g. the CNOT at node A and the X gate at node B, etc.) can be executed at the right time, i.e. only after the entangled pair has been acquired. Subsequent transfer of measurement results (classical bits) can be done via the (classical) Internet as the needed by the operations – synchronization here is for a node to wait for the bits from the other node before proceeding. (Note that a quantum circuit might be executed a number of times to collect statistics – not quite necessary for the simple distributed CNOT example here.) Second, the status of execution and, where required, results are returned from the nodes to the controller. One can generalize the above to more than two nodes and more complicated circuits, but the key point here is the access to the quantum Internet and classical Internet at different points of the circuit, with the required synchronization among nodes. Moreover, a large circuit (with a large depth) might be executed in stages (not shown in Figure 4.7), i.e. a part of the quantum circuit executed and then back to classical control, and then the next part of the quantum circuit executed and back to classical control, and so on, that is, greater interleaving between classical control using classical computer(s) and quantum execution on the quantum computer(s).

A more detailed discussion of required control and scheduling of programs in distributed quantum computations is in Parekh et al. [2021]. A reference model for distributed quantum computation called D-NISQ is presented in Acampora et al. [2023], where a large problem might be decomposed into subproblems to be solved on multiple quantum processors.

---

3 The program executed on each node can be viewed as a *hybrid classical-quantum program*, i.e. a program containing classical control instructions and quantum operations to be executed; see Dahlberg et al. [2022] for details, and an example at https://github.com/brunorijsman/quantum-internet-hackathon-2022/blob/main/docs/qne-adk-implementation.md [last accessed: 8/3/2023] though in the context of a simulated distributed environment NetQASM.

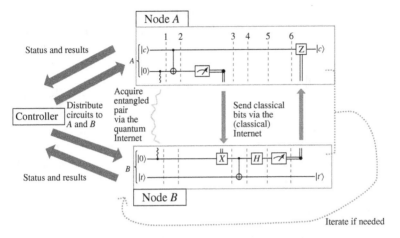

Figure 4.7 An illustration of execution of a four-qubit circuit for the distributed CNOT.

Note that timing is a factor since quantum states can *decohere* over time. As mentioned in Vardoyan et al. [2022], a quantum state (represented by density operator ρ_0) placed into quantum memory at time $t = 0$ might be subjected to noise (e.g. noise in the environment due to imperfect shielding) so that after time t, the state might be ρ_t. There could also be noise due to imperfect quantum gate implementations. There are a number of different mathematical models of noise, in the form of *completely positive trace preserving* (CPTP) maps, one of which is the following map $\mathcal{D}_t$:

$$\rho_t = \mathcal{D}_t(\rho) = (1 - 3p)\rho + pX\rho X + pY\rho Y + pZ\rho Z = (1 - 4p)\rho + 4p\frac{\mathbf{I}}{2}$$

where p is a probability and using the fact that $\frac{1}{4}(\rho + X\rho X + Z\rho Z + XZ\rho(XZ)^\dagger) = \frac{1}{2}\mathbf{I}$ and $Y = iXZ$ (which we will show in Chapter 5). But the idea of this map is that if p is high, then the quantum state can approach the *maximally mixed state* $\frac{\mathbf{I}}{2}$. The time dependency of p can be expressed by setting $p = \frac{1}{4}(1 - e^{-\frac{t}{T}})$, for some constant T, where it can be seen that p increases over time. There are other noise models such as the dephasing and damping noise models which we will not go into here. This means that suppose it takes too long to receive certain required bits or even to get an entangled state, then the quantum state might have already decohered while waiting! This also means that the depth of the circuits to be executed can affect the chance of completion of the operations. Another important parameter is entanglement fidelity, where the "quality" of the entanglement might be affected by noise (we talk further about fidelity in Chapter 6).

Architectures for distributed quantum computers is still a research area. A point to note is that, on each node, there would be qubits used for computations and qubits used for entanglement (with other nodes), so called *computation qubits*, and *communication qubits* (e.g. the qubits involved in an entangled pair across two nodes, with one qubit on each node, which is also called an *entanglement link* in this context as it "links" the two nodes). An extensive study of two different quantum architectures for coupling the networking and computing components is given in Vardoyan et al. [2022].

Many of the underlying details of control and scheduling of quantum operations, which might vary according to the actual implementation of quits and quantum gates, should be abstracted away from programmers. An interesting development is the QMPI [Häner et al., 2021] is a quantum extension of the classical MPI (Message Passing Interface) that provides a device-independent library for writing distributed computing programs. For example, a program using QMPI might use library calls to acquire quantum resources, e.g. to request an EPR pair to be generated between two nodes, each node invokes the function call:

```
QMPI_Prepare_EPR(qubit, dest, tag, comm)
```

where qubit is a qubit in |0⟩, dest refers to the destination of a QMPI process running on the other node, tag is a message tag, and comm is the communicator (referring to a collection of (quantum) nodes). Note that there is also an asynchronous version of the call allowing EPR pairs to be acquired ahead of time (of course, this is provided the EPR pairs can remain in memory long enough before they decohere!). Library operations such as QMPI_Send and QMPI_Recv enable teleporting of qubits, and there are also collective operations for multipartite operations, for example, the GHZ state can be coded using QMPI, together with the corresponding operations for uncomputing (reversing computations).

4.5 Distributed Quantum Computing Without Internode Entanglement

So far, we have seen how entanglement between nodes (between qubits on the nodes) can be used to compute quantum circuits distributed across nodes. But today (at least at the time of writing), we have quantum computers (offered via cloud services) that don't communicate quantumly, i.e. no entanglement among two computers. Would it still be useful to harness multiple quantum computers which are not connected via entanglement?

The work by Avron et al. [2021] showed that instead of using one n-qubit circuit (or quantum computer), it might be better to use two $(n-1)$-qubit circuits (or quantum computers) to compute an equivalent solution for a particular type of function. This means that $2(n-1)$ qubits are used instead of n qubits but the depth of the $(n-1)$-qubit circuits could be much lower than the n-qubit quantum circuit for the equivalent computation. Lower depth of a circuit implies a noise reduction (since noise can build up with depth – in fact, the noise in the output can grow exponentially with the depth of the circuit).

Consider a function f of the form $f : \{0,1\}^n \to \{0,1\}^m$ for $m \leq n-1$. This function can be equivalently specified as two functions $f_{even/odd} : \{0,1\}^{n-1} \to \{0,1\}^m$ defined as follows:

$$f_{even}(y) = f_{even}(y_1...y_{n-1}) = f(0y_1...y_{n-1}) = f(0y), \text{ and}$$
$$f_{odd}(y) = f_{odd}(y_1...y_{n-1}) = f(1y_1...y_{n-1}) = f(1y)$$

where $y \in \{0,1\}^{n-1}$. So, to compute f on some n-bit string, first look at the first bit, which will then determine which of f_{even} or f_{odd} to use on the rest of the bits.

Useful for Grover and Deutsch–Jozsa algorithms (and perhaps others) would be a quantum circuit that can compute with n-qubits, denoted with $|x\rangle$ where $x \in \{0,1\}^n$: $U_f |x\rangle = (-1)^{f(x)} |x\rangle$. But assuming we have $(n-1)$-qubit circuits for computing the following: $U_{f_{even}} |y\rangle = (-1)^{f_{even}(y)} |y\rangle$ and $U_{f_{odd}} |y\rangle = (-1)^{f_{odd}(y)} |y\rangle$, where $y \in \{0,1\}^{n-1}$, one can see that:

$$U_f |0\rangle \otimes |y\rangle = (-1)^{f(0y)} |0\rangle \otimes |y\rangle = |0\rangle \otimes (-1)^{f_{even}(y)} |y\rangle = |0\rangle \otimes U_{f_{even}} |y\rangle$$
$$\text{and: } U_f |1\rangle \otimes |y\rangle = (-1)^{f(1y)} |1\rangle \otimes |y\rangle = |1\rangle \otimes (-1)^{f_{odd}(y)} |y\rangle = |1\rangle \otimes U_{f_{odd}} |y\rangle$$

that is, we can compute U_f with the help of $U_{f_{odd}}$ and $U_{f_{even}}$, and though we use more qubits doing so, because the two circuits for $U_{f_{odd}}$ and $U_{f_{even}}$ can be much lower in depth than U_f, we can effectively get the results with reduced noise. This can perhaps be generalized to more nodes with a similar decomposition of $U_{f_{odd}}$ and $U_{f_{even}}$ into their corresponding "even" and "odd' functions and continuing recursively.

4.6 Summary

We have looked at some of the fundamental ideas in implementing distributed quantum computations. Distributed quantum computing continues to be an active area of research and involves many different aspects including hardware architectures for quantum computers that can work with other quantum computers over a network, methods to partition a large quantum circuit to be computed efficiently over a collection of quantum nodes, control, and scheduling mechanisms that are sufficiently efficient, and programming models and abstractions. There are now libraries such as QMPI conceived to help programmers, tested at least in simulations. The future could see more sophisticated libraries, e.g. implementing some of the quantum protocols we saw in Chapter 3. Also, clearly, research into quantum architectures and ways to store and use qubits more reliably (e.g. able to last for longer times) would be beneficial.

References

Giovanni Acampora, Ferdinando Di Martino, Alfredo Massa, Roberto Schiattarella, and Autilia Vitiello. D-NISQ: a reference model for distributed noisy intermediate-scale quantum computers. *Information Fusion*, 89:16–28, 2023.

Pablo Andrés-Martínez and Chris Heunen. Automated distribution of quantum circuits via hypergraph partitioning. *Physical Review A*, 100:032308, Sep 2019. doi: 10.1103/PhysRevA.100.032308.

J. Avron, Ofer Casper, and Ilan Rozen. Quantum advantage and noise reduction in distributed quantum computing. *Physical Review A*, 104:052404, Nov 2021. doi: 10.1103/PhysRevA.104.052404.

Sebastian de Bone, Runsheng Ouyang, Kenneth Goodenough, and David Elkouss. Protocols for creating and distilling multipartite GHZ states with Bell pairs. *IEEE Transactions on Quantum Engineering*, 1:1–10, 2020. doi: 10.1109/TQE.2020 .3044179.

Daniele Cuomo, Marcello Caleffi, Kevin Krsulich, Filippo Tramonto, Gabriele Agliardi, Enrico Prati, and Angela Sara Cacciapuoti. Optimized compiler for distributed quantum computing. *ACM Transactions on Quantum Computing*, 4(2), Feb 2023. ISSN 2643-6809. doi: 10.1145/3579367.

Axel Dahlberg, Bart van der Vecht, Carlo Delle Donne, Matthew Skrzypczyk, Ingmar te Raa, Wojciech Kozlowski, and Stephanie Wehner. NetQASM—a low-level instruction set architecture for hybrid quantum-classical programs in a quantum internet. *Quantum Science and Technology*, 7(3):035023, Jun 2022. doi: 10.1088/2058-9565/ac753f.

Zohreh Davarzani, Mariam Zomorodi, and Mahboobeh Houshmand. A hierarchical approach for building distributed quantum systems. *Scientific Reports*, 12(1):15421, 2022.

J. Eisert, K. Jacobs, P. Papadopoulos, and M. B. Plenio. Optimal local implementation of nonlocal quantum gates. *Physical Review A*, 62:052317, Oct 2000. doi: 10.1103/PhysRevA.62.052317.

Davide Ferrari, Angela Sara Cacciapuoti, Michele Amoretti, and Marcello Caleffi. Compiler design for distributed quantum computing. *IEEE Transactions on Quantum Engineering*, 2:1–20, 2021. doi: 10.1109/TQE.2021.3053921.

Thomas Häner, Damian S. Steiger, Torsten Hoefler, and Matthias Troyer. Distributed quantum computing with QMPI. In *Proceedings of the International Conference for High Performance Computing, Networking, Storage and Analysis, SC '21*, New York, NY, USA, 2021. Association for Computing Machinery. ISBN 9781450384421. doi: 10.1145/3458817.3476172.

R. van Houte, J. Mulderij, T. Attema, I. Chiscop, and F. Phillipson. Mathematical formulation of quantum circuit design problems in networks of quantum computers. *Quantum Information Processing*, 19(5):141, Mar 2020. ISSN 1570-0755. doi: 10.1007/s11128-020-02630-8.

Yunseok Kwak, Won Joon Yun, Jae Pyoung Kim, Hyunhee Cho, Jihong Park, Minseok Choi, Soyi Jung, and Joongheon Kim. Quantum distributed deep learning architectures: models, discussions, and applications. *ICT Express*, 9(3):486–491, 2022.

Niels M. P. Neumann and Robert S. Wezeman. Distributed quantum machine learning, page 8, 2022. doi: 10.1007/978-3-031-06668-9_20.

R. Parekh, A. Ricciardi, A. Darwish, and S. DiAdamo. Quantum algorithms and simulation for parallel and distributed quantum computing. In *2021 IEEE/ACM Second International Workshop on Quantum Computing Software (QCS)*, pages 9–19, Los Alamitos, CA, USA, Nov 2021. IEEE Computer Society. doi: 10.1109/QCS54837.2021.00005. URL https://doi.ieeecomputersociety.org/10.1109/QCS54837.2021.00005.

Ranjani G. Sundaram, Himanshu Gupta, and C. R. Ramakrishnan. Efficient distribution of quantum circuits. In Seth Gilbert, editor, *35th International Symposium on Distributed Computing, DISC 2021*, October 4–8, 2021, Freiburg, Germany (Virtual Conference), volume 209 of LIPIcs, pages 41:1–41:20. Schloss Dagstuhl–Leibniz-Zentrum für Informatik, 2021. doi: 10.4230/LIPIcs.DISC.2021.41.

Jiawei Tan, Ligang Xiao, Daowen Qiu, Le Luo, and Paulo Mateus. Distributed quantum algorithm for Simon's problem. *Physical Review A*, 106:032417, Sep 2022. doi: 10.1103/PhysRevA.106.032417.

Wei Tang, Teague Tomesh, Martin Suchara, Jeffrey Larson, and Margaret Martonosi. CutQC: using small quantum computers for large quantum circuit evaluations. In *Proceedings of the 26th ACM International Conference on Architectural Support for Programming Languages and Operating Systems, ASPLOS '21*, pages 473–486, New York, NY, USA, 2021. Association for Computing Machinery. ISBN 9781450383172. doi: 10.1145/3445814.3446758.

Gayane Vardoyan, Matthew Skrzypczyk, and Stephanie Wehner. On the quantum performance evaluation of two distributed quantum architectures. *Performance Evaluation*, 153:102242, 2022.

Anocha Yimsiriwattana and Samuel J. Lomonaco Jr. Distributed quantum computing: a distributed Shor algorithm. In Eric Donkor, Andrew R. Pirich, and Howard E. Brandt, editors, *Quantum Information and Computation II*, volume 5436, pages 360–372. International Society for Optics and Photonics, SPIE, 2004.

Anocha Yimsiriwattana and Samuel J. Lomonaco Jr. Generalized GHZ states and distributed quantum computing. In *Coding Theory and Quantum Computing, Contemporary Mathematics*, pages 131–147. AMS, Providence, RI, 2005. Also available at https://arxiv.org/abs/quant-ph/0402148.

5

Delegating Quantum Computations

This chapter discusses the idea of a client, with limited resources, delegating quantum computations to a more powerful quantum computer or server, in the style of what we see with cloud computing today. However, security and privacy are key considerations when delegating computations. We look at how to delegate quantum computations to a quantum server in such a way as to protect the privacy of the client's inputs, that is, the server cannot know the client's inputs while still being able to perform the computations for the client. The chapter also looks at how the client can verify the server's computations, or at least detect if the server really performed the delegated computations correctly. Finally, the chapter looks briefly at the idea of quantum computing as a service (QaaS).

5.1 Delegating Private Quantum Computations

This section considers the idea of private computations or data that the client wants hidden from the server but still wanting the server to help it perform those computations. This might seem to be impossible, but has been possible in the classical case with a technique such as homomorphic encryption,[1] where data can be computed with while still in the encrypted form.

We consider the techniques mainly by Broadbent [2015] and Fisher et al. [2014], though acknowledge the pioneering work of Childs [2005], in computing over encrypted quantum data.

We first introduce the idea of the *classical one-time pad* and the *quantum one-time pad*. The classical one-time pad is a procedure for encrypting a bit string $x = x_1 \ldots x_n$ by mapping each bit x_i to $x_i \oplus r_i$ for some random key bit-string $r = r_1 \ldots r_n \in \{0,1\}^n$. Since each $x_i \oplus r_i$ is uniformly random given that r is secret, x is concealed. Analogous to the classical one-time pad is the quantum

---

1 See https://homomorphicencryption.org/introduction [last accessed: 14/11/2022].

From Distributed Quantum Computing to Quantum Internet Computing: An Introduction, First Edition. Seng W. Loke.

one-time pad which effectively encrypts a quantum state $|x\rangle \langle x| = \rho$ by applying an operator that is chosen uniformly randomly from the set $\{\mathbf{I}, \mathbf{X}, \mathbf{Z}, \mathbf{XZ}\}$, that is, ρ is mapped to:

$$\frac{1}{4} \left(\rho + X\rho X + Z\rho Z + XZ\rho (XZ)^{\dagger} \right) = \frac{1}{4} \sum_{a,b \in \{0,1\}} X^a Z^b \rho Z^b X^a = \frac{1}{2} \mathbf{I}$$

which is the maximally mixed state so that the state is concealed.[2] To see that $\frac{1}{4} \sum_{a,b \in \{0,1\}} X^a Z^b \rho X^a Z^b = \frac{1}{2} \mathbf{I}$, note that the Pauli matrices form a basis for the space of 2×2 complex matrices, and so, for some α, β, γ, and δ, we can write ρ as follows:

$$\rho = \alpha \mathbf{I} + \beta X + \gamma Y + \delta Z$$

and observe that, since $XX = \mathbf{I}$, $ZZ = \mathbf{I}$, $YY = \mathbf{I}$, $XZ = -ZX$, and $Y = iXZ$, we have:

$$X\rho X = \alpha \mathbf{I} + \beta XXX + \gamma XYX + \delta XZX = \alpha \mathbf{I} + \beta X - \gamma Y - \delta Z$$

$$Z\rho Z = \alpha \mathbf{I} + \beta ZXZ + \gamma ZYZ + \delta ZZZ = \alpha \mathbf{I} - \beta X - \gamma Y + \delta Z$$

$$XZ\rho ZX = \alpha \mathbf{I} + \beta XZXZX + \gamma XZYZX + \delta XZZZX = \alpha \mathbf{I} - \beta X + \gamma Y - \delta Z$$

and $\alpha = 1/2$ (since $tr(\rho) = tr(\alpha \mathbf{I} + \beta X + \gamma Y + \delta Z) = \alpha \cdot tr(\mathbf{I}) + \beta \cdot tr(X) + \gamma \cdot tr(Y) + \delta \cdot tr(Z) = \alpha \cdot 2 = 1$, by the linearity of trace), we have (using the above equations):

$$\frac{1}{4} \sum_{a,b \in \{0,1\}} X^a Z^b \rho Z^b X^a = \frac{1}{4} (\rho + X\rho X + Z\rho Z + XZ\rho ZX)$$

$$= \frac{1}{4} (\alpha \mathbf{I} + \alpha \mathbf{I} + \alpha \mathbf{I} + \alpha \mathbf{I}) = \frac{1}{2} \mathbf{I}$$

Suppose ρ' is the density operator for an n-qubit state, one can encrypt each qubit, and the above generalizes to n-qubits as follows:

$$\frac{1}{4^n} \sum_{\mathbf{a},\mathbf{b} \in \{0,1\}^n} \left(X^{a_1} Z^{b_1} \otimes \cdots \otimes X^{a_n} Z^{b_n} \right) \rho' \left(Z^{b_1} X^{a_1} \otimes \cdots \otimes Z^{b_n} X^{a_n} \right) = \frac{\mathbf{I}}{2^n}$$

where $\mathbf{a} = a_1 \ldots a_n$ and $\mathbf{b} = b_1 \ldots b_n$ are n-bit vectors. The LHS can be rewritten as

$$\frac{1}{4^n} \sum_{\mathbf{a},\mathbf{b} \in \{0,1\}^n} (X^{a_1} \otimes \cdots \otimes X^{a_n})(Z^{b_1} \otimes \cdots \otimes Z^{b_n})\rho'(Z^{b_1} \otimes \cdots \otimes Z^{b_n})(X^{a_1} \otimes \cdots \otimes X^{a_n})$$

---

2 One way to look at a maximally mixed state is that it is a state with minimal information. For example, writing:

$$\frac{1}{2} \mathbf{I} = \frac{1}{2} |0\rangle \langle 0| + \frac{1}{2} |1\rangle \langle 1|$$

we see that there is equal probability of $|0\rangle \langle 0|$ and $|1\rangle \langle 1|$, or one can look at this information theoretically and use the *von Neumann entropy* of a quantum state ρ defined by $-tr(\rho \log\rho)$: the von Neumann entropy of $\frac{1}{2} \mathbf{I}$ is $-tr(\frac{1}{2} \mathbf{I} \log(\frac{1}{2}\mathbf{I})) = \log 2 = 1$ (taking log to base 2, and taking $\log(\sum_a c_a |a\rangle \langle a|) = \sum_a \log(c_a) |a\rangle \langle a|$, based on the idea that given a diagonal matrix A with diagonal elements A_1 to A_n, f(A) denotes a diagonal matrix with diagonal elements $f(A_1)$ to $f(A_n)$.)

that is, we have: $\frac{1}{4^n} \sum_{\mathbf{a},\mathbf{b} \in \{0,1\}^n} \mathbf{X}(\mathbf{a}) \mathbf{Z}(\mathbf{b}) \rho' \mathbf{Z}(\mathbf{b}) \mathbf{X}(\mathbf{a}) = \frac{1}{2^n} \mathbf{I}$, where $\mathbf{X}(\mathbf{a}) = X^{a_1} \otimes \cdots \otimes X^{a_n}$ and $\mathbf{Z}(\mathbf{b}) = Z^{b_1} \otimes \cdots \otimes Z^{b_n}$.

Before proceeding, it is important to note the limitations we are imposing on the client. We assume that the client A is not capable of universal quantum computation (and so is using the powerful server B for this purpose where B can do universal quantum computation, i.e. perform a range of different gate operations as we see later), but A can do the following:

- *Assumption 1*: Prepare and send to B qubits in states randomly chosen from a set of limited possibilities, e.g. to be able to prepare states equivalent to the set $\{|+\rangle = \frac{1}{\sqrt{2}}(|0\rangle + |1\rangle), |-\rangle = \frac{1}{\sqrt{2}}(|0\rangle - |1\rangle), |+_y\rangle = \frac{1}{\sqrt{2}}(|0\rangle + i|1\rangle), |-_y\rangle = \frac{1}{\sqrt{2}}(|0\rangle - i|1\rangle)\}$ and receive qubits from B, and
- *Assumption 2*: Can apply single-qubit Pauli operators (i.e. X and Z) in order to encrypt qubits to be sent to B and to decrypt output qubits received from B.

Consider an example of the use of the quantum one-time pad. A has $|\psi\rangle$ and then sends an encrypted quantum state to B (e.g. it actually sends $|\psi'\rangle = XZ|\psi\rangle$ without telling B that it is using XZ), to get B to apply operations on the state. B sees the state as effectively $\frac{1}{2}\mathbf{I}$ since it does not know which operation A used, and so, cannot know what really $|\psi\rangle$ is, even with measurements. Now, suppose B is to help A perform an X gate on $|\psi\rangle$, then on receiving $|\psi'\rangle$, B performs X on it to get $X|\psi'\rangle$ and sends this back to A. A, on receiving $X|\psi'\rangle$ then applies ZX to it to get:

$$ZX(X|\psi'\rangle) = ZX(X(XZ|\psi\rangle)) = ZXXXZ|\psi\rangle = -X|\psi\rangle$$

since $XX = ZZ = \mathbf{I}$ and $ZX + XZ = 0$, and we can safely ignore the global phase "$-$," i.e. A then gets $X|\psi\rangle$ which is what it wanted. In general, in the case of the X gate operation, the decryption key for encryption using the operation $X^a Z^b$ is to apply $Z^b X^a$ since

$$Z^b X^a X X^a Z^b = Z^b X Z^b = \begin{cases} -X, & b = 1 \\ X, & b = 0 \end{cases}$$

What if A wanted B to perform a Z gate instead of the X gate? A has $|\psi\rangle$ and then sends an encrypted quantum state to B, i.e. choosing $a, b \in \{0, 1\}$ it sends $|\psi'\rangle = X^a Z^b |\psi\rangle$ without telling B its choice of a and b, to get B to apply Z on the state. B performs Z on it to get $Z|\psi'\rangle$ and sends this back to A. A, on receiving $Z|\psi'\rangle$, then applies $Z^b X^a$ (similar to the case of the X gate) to it to get:

$$Z^b X^a \left(Z|\psi'\rangle \right) = Z^b X^a Z X^a Z^b |\psi\rangle = \begin{cases} -Z|\psi\rangle, & a = 1 \\ Z|\psi\rangle, & a = 0 \end{cases}$$

Ignoring the global phase "$-$," A then gets $Z|\psi\rangle$ as required.

What if A wanted B to perform a H gate instead of a Z gate? As before, A has $|\psi\rangle$ and then sends an encrypted quantum state to B, i.e. choosing $a, b \in \{0, 1\}$ it sends $|\psi'\rangle = X^a Z^b |\psi\rangle$, to get B to apply H on the state. B performs H on it to get $H |\psi'\rangle$ and sends this back to A. A, on receiving $H |\psi'\rangle$, then applies $Z^a X^b$ to it (note that this decryption operation is different from $Z^b X^a$ used for the X and Z gates) to get:

$$Z^a X^b \left(H |\psi'\rangle\right) = Z^a X^b H X^a Z^b |\psi\rangle$$
$$= Z^a X^b Z^a H Z^b |\psi\rangle \quad \text{(since } H X^a = Z^a H\text{)}$$
$$= Z^a X^b Z^a X^b H |\psi\rangle \quad \text{(since } H Z^b = X^b H\text{)}$$

$$= \begin{cases} H |\psi\rangle, & \text{if } a = 0, b = 0 \\ XXH |\psi\rangle = H |\psi\rangle, & \text{if } a = 0, b = 1 \\ ZZH |\psi\rangle = H |\psi\rangle, & \text{if } a = 1, b = 0 \\ ZXZXH |\psi\rangle = -ZXXZH |\psi\rangle = -H |\psi\rangle, & \text{if } a = 1, b = 1 \end{cases}$$

Ignoring the global phase "$-$," A then gets $H |\psi\rangle$ as required.

What if A wanted B to perform a P gate instead of a Z gate? The P gate or phase gate takes the following form: $\begin{bmatrix} 1 & 0 \\ 0 & i \end{bmatrix}$, that is, $P |0\rangle = |0\rangle$, and $P |1\rangle = i |1\rangle$. As before, A has $|\psi\rangle$ and then sends an encrypted quantum state to B, i.e. choosing $a, b \in \{0, 1\}$ it sends $|\psi'\rangle = X^a Z^b |\psi\rangle$, to get B to apply P on the state. B performs P on it to get $P |\psi'\rangle$ and sends this back to A. A, on receiving $P |\psi'\rangle$, then applies $Z^{a \oplus b} X^a$ to it (note this is different from the cases above) to get:

$$Z^{a \oplus b} X^a P |\psi'\rangle = Z^{a \oplus b} X^a P X^a Z^b |\psi\rangle$$
$$= (-i)^a Z^{a \oplus b} X^a Z^a X^a P Z^b |\psi\rangle \quad \text{(since } P X^a = (-i)^a Z^a X^a P\text{)}$$
$$= (-i)^a Z^{a \oplus b} X^a Z^a X^a Z^b P |\psi\rangle \quad \text{(since } P Z^b = Z^b P\text{)}$$

$$= \begin{cases} P |\psi\rangle, & \text{if } a = 0, \ b = 0 \\ ZZP |\psi\rangle = P |\psi\rangle, & \text{if } a = 0, \ b = 1 \\ (-i)ZXZXP |\psi\rangle = iZXXZP |\psi\rangle = iP |\psi\rangle, & \text{if } a = 1, \ b = 0 \\ (-i)XZXZP |\psi\rangle = iZXXZP |\psi\rangle = iP |\psi\rangle, & \text{if } a = 1, \ b = 1 \end{cases}$$

Ignoring the global phase "i," A then gets $P |\psi\rangle$ as required.

What if A wanted B to perform a *CNOT* gate instead? A now has a two qubit state $|\psi\rangle$ and then sends the encrypted quantum state to B, i.e. choosing $a, b, c, d \in \{0, 1\}$ it sends $|\psi'\rangle = X^a Z^b \otimes X^c Z^d |\psi\rangle$, to get B to apply *CNOT* on the state. B performs *CNOT* on it to get *CNOT* $|\psi'\rangle$ and sends this back to A. A, on receiving *CNOT* $|\psi'\rangle$, then applies $Z^{b+d} X^a \otimes Z^d X^{a+c}$ to it to get:

$(Z^{b+d}X^a \otimes Z^d X^{a+c})CNOT|\psi'\rangle$

$= (Z^{b+d}X^a \otimes Z^d X^{a+c})CNOT((X^a Z^b \otimes X^c Z^d)|\psi\rangle)$

$= (Z^{b+d}X^a \otimes Z^d X^{a+c})CNOT(X^a \otimes X^c)(Z^b \otimes Z^d)|\psi\rangle$

$= (Z^{b+d}X^a \otimes Z^d X^{a+c})CNOT(X^a \otimes I)(I \otimes X^c)(Z^b \otimes I)(I \otimes Z^d)|\psi\rangle$

$= (Z^{b+d}X^a \otimes Z^d X^{a+c})(X^a \otimes X^a)(I \otimes X^c)CNOT(Z^b \otimes I)(I \otimes Z^d)|\psi\rangle$

（since $CNOT(X^a \otimes I) = (X^a \otimes X^a)CNOT$ and $CNOT(I \otimes X^c)$

$= (I \otimes X^c)CNOT)$

$= (Z^{b+d}X^a \otimes Z^d X^{a+c})(X^a \otimes X^a)(I \otimes X^c)(Z^b \otimes I)(Z^d \otimes Z^d)CNOT|\psi\rangle$

（since $CNOT(Z^b \otimes I) = (Z^b \otimes I)CNOT$ and $CNOT(I \otimes Z^d)$

$= (Z^d \otimes Z^d)CNOT)$

$= (Z^{b+d}X^a X^a Z^b Z^d \otimes Z^d X^{a+c} X^a X^c Z^d)CNOT|\psi\rangle = CNOT|\psi\rangle$

where we have repeatedly used the identity $AB \otimes CD = (A \otimes C)(B \otimes D)$. So, A gets $CNOT|\psi\rangle$ as required. Note that we could use $Z^{b\oplus d}X^a \otimes Z^d X^{a\oplus c}$ as well with the same outcome, replacing "+" with "⊕."

What if A wanted B to perform a R gate instead? Note that the R gate is $\begin{bmatrix} 1 & 0 \\ 0 & e^{i\pi/4} \end{bmatrix}$, that is, $R|0\rangle = |0\rangle$, and $R|1\rangle = e^{i\pi/4}|1\rangle$. What if we used the approach we have been using with an appropriate operation to recover the result? Suppose A now has a one-qubit state $|\psi\rangle$ and then sends the encrypted quantum state to B, i.e. choosing $a, b \in \{0, 1\}$ it sends $|\psi'\rangle = X^a Z^b |\psi\rangle$, to get B to apply R on the state. B then sends back $R|\psi'\rangle$ to A. How can A recover the $R|\psi\rangle$ from $R|\psi'\rangle$? As noted in Broadbent [2015], this is not so straightforward since we have:

$R|\psi'\rangle = RX^a Z^b |\psi\rangle$

$= \begin{cases} R|\psi\rangle, & \text{if } a = 0, \ b = 0 \\ RZ|\psi\rangle = ZR|\psi\rangle, & \text{if } a = 0, \ b = 1 \\ RX|\psi\rangle = XZPR|\psi\rangle, & \text{if } a = 1, \ b = 0 \\ RXZ|\psi\rangle = XZPRZ|\psi\rangle = XZPZR|\psi\rangle \\ = XPZZR|\psi\rangle = XPR|\psi\rangle, & \text{if } a = 1, \ b = 1 \end{cases}$

since $RZ = ZR$, $RX = XZPR$ (ignoring global phases), and $ZP = PZ$, that is, $R|\psi'\rangle = X^a Z^{a\oplus b} P^a R|\psi\rangle$. This means that when $a = 0$, A can just apply Z^b to $R|\psi'\rangle$ to obtain $R|\psi\rangle$: $Z^b(R|\psi'\rangle) = Z^b(Z^b R|\psi\rangle) = R|\psi\rangle$. But when $a = 1$, we have the additional P in front of the R. One needs to apply ZP to remove the P, e.g. when $a = 1$, $b = 1$, one can apply ZPX to $XPR|\psi\rangle$ to get $ZPX(XPR|\psi\rangle) = ZPPR|\psi\rangle = ZZR|\psi\rangle = R|\psi\rangle$, since $Z = PP$. However, recall that we have assumed that the client is limited and cannot perform P by itself (and so, it is using the more powerful B to do so!), and so cannot perform ZP

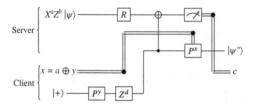

Figure 5.1 Protocol for an R-gate, where the output from the server is $|\psi''\rangle = X^{a\oplus c}Z^{a\cdot(c\oplus y\oplus 1)\oplus b\oplus d\oplus y}R\,|\psi\rangle$. The client chooses randomly and uniformly $y, d \in \{0,1\}$. This is a way to get the server to perform the P operation in a way controlled by the client. This can be called an R-gate "gadget."

to remove the P. This can be solved in two ways. One way is by asking B to do this additional P operation, as we have seen earlier, but this needs to be done when $a = 1$ as well as when $a = 0$ (using a dummy qubit in the case $a = 0$, so that B does not know whether $a = 1$), which involves two-way quantum communication.

The other way is using the following protocol from Broadbent [2015], which we outline below, which sends, in addition to $X^a Z^b\,|\psi\rangle$, an auxiliary qubit in a state randomly chosen out of a set of four possible ones, and a classical bit, as explained below. The server sends back the result and a classical bit to decrypt the result. The circuit to implement this protocol is shown in Figure 5.1.

The client, say A, chooses randomly and uniformly $a, b \in \{0,1\}$ to encrypt the state $|\psi\rangle$ as $X^a Z^b\,|\psi\rangle$. A sends $X^a Z^b\,|\psi\rangle$ to the server, i.e. B. In addition, A chooses randomly and uniformly $y, d \in \{0,1\}$, computes $x = a \oplus y$, and sends x and the quantum state $Z^d P^y\,|+\rangle$ to the server B. This is a way to get the server to perform the P operation in a way controlled by the client, without revealing any information to the server about a (remember we mentioned above that when $a = 1$, we need to apply an additional P and this circuit makes the server does so – to see this, when $a = 0, x = a \oplus y = y$, and looking at the gates along the quantum wire in Figure 5.1 from $|+\rangle$ to $|\psi''\rangle$, when $x = y = 1$, we have $P^x Z^d P^y = PZ^d P = PPZ^d = ZZ^d$, and when $x = y = 0$, we have $P^x Z^d P^y = Z^d$, so that no P is applied, but when $a = 1$, $x = 1 \oplus y$, i.e. we have either $x = 0$ and $y = 1$ or $x = 1$ and $y = 0$, which means that $P^x Z^d P^y = PZ^d$ or $P^x Z^d P^y = Z^d P$, so that the operation P is applied; note that the server, even when knowing x, does not know what a is since y can be 0 or 1).

Note that the client A does not actually apply P to $|+\rangle$ when $y = 1$, but as in our assumptions, we assume that the client can choose randomly and uniformly from one of the following states: $\{|+\rangle = \frac{1}{\sqrt{2}}(|0\rangle + |1\rangle), |-\rangle = \frac{1}{\sqrt{2}}(|0\rangle - |1\rangle), |+_y\rangle = \frac{1}{\sqrt{2}}(|0\rangle + i\,|1\rangle), |-_y\rangle = \frac{1}{\sqrt{2}}(|0\rangle - i\,|1\rangle)\}$ (our Assumption 1 stated earlier), which is equivalent to the client choosing randomly and uniformly $y, d \in \{0,1\}$ and computing $Z^d P^y\,|+\rangle$ since:

y	d	$Z^d P^y \, |+\rangle$
0	0	$|+\rangle$
0	1	$Z \, |+\rangle = |-\rangle$
1	0	$P \, |+\rangle = \frac{1}{\sqrt{2}}(|0\rangle + i \, |1\rangle) = \left|+_y\right\rangle$
1	1	$ZP \, |+\rangle = \frac{1}{\sqrt{2}}(|0\rangle - i \, |1\rangle) = \left|-_y\right\rangle$

On receiving the input $X^a Z^b \, |\psi\rangle$, x and $Z^d P^y \, |+\rangle$ from the client (i.e. A), the server (i.e. B) applies the R operator and does a CNOT with control $Z^d P^y \, |+\rangle$ and then performs P^x, depending on x, and measures the target; the measured outcome c is sent back to the client A. From the protocol, the server computes $|\psi''\rangle$ given by

$$|\psi''\rangle = X^{a \oplus c} Z^{a \cdot (c \oplus y \oplus 1) \oplus b \oplus d \oplus y} R \, |\psi\rangle$$

To see why this is so, the circuit in Figure 5.1 can be derived as follows, from the Appendix in Broadbent [2015]:

(1) We start with the following circuit which swaps $|+\rangle$ and $|\psi\rangle$.

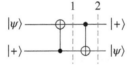

The circuit can be shown to do the swap as follows. Let $|\psi\rangle = \alpha \, |0\rangle + \beta \, |1\rangle$. Writing the lower qubit first, the circuit starts with:

$$|+\rangle \, |\psi\rangle = |+\rangle \, (\alpha \, |0\rangle + \beta \, |1\rangle) = \frac{1}{\sqrt{2}}(|0\rangle + |1\rangle)(\alpha \, |0\rangle + \beta \, |1\rangle)$$

$$= \frac{1}{\sqrt{2}}(\alpha \, |00\rangle + \beta \, |01\rangle + \alpha \, |10\rangle + \beta \, |11\rangle)$$

Then, applying the first CNOT, we have at slice 1:

$$\frac{1}{\sqrt{2}}(\alpha \, |00\rangle + \beta \, |01\rangle + \alpha \, |11\rangle + \beta \, |10\rangle)$$

And then, we apply the second CNOT to get at slice 2:

$$\frac{1}{\sqrt{2}}(\alpha \, |00\rangle + \beta \, |11\rangle + \alpha \, |01\rangle + \beta \, |10\rangle)$$

$$= \frac{1}{\sqrt{2}}((\alpha \, |0\rangle + \beta \, |1\rangle) \, |1\rangle + (\alpha \, |0\rangle + \beta \, |1\rangle) \, |0\rangle)$$

$$= \frac{1}{\sqrt{2}}(|\psi\rangle \, |1\rangle + |\psi\rangle \, |0\rangle) = |\psi\rangle \, |+\rangle$$

(2) We observe from above that, after the first CNOT, we have: $\frac{1}{\sqrt{2}}(\alpha\,|00\rangle +$ $\beta\,|01\rangle + \alpha\,|11\rangle + \beta\,|10\rangle)$, which can be rewritten as $\frac{1}{\sqrt{2}}((\alpha\,|0\rangle + \beta\,|1\rangle)\,|0\rangle +$ $(\alpha\,|1\rangle + \beta\,|0\rangle)\,|1\rangle) = \frac{1}{\sqrt{2}}(|\psi\rangle\,|0\rangle + X\,|\psi\rangle\,|1\rangle)$. So, if we measure the top qubit (the qubit on the right), we see that if the measurement reveals $|0\rangle$, the left/lower qubit is $|\psi\rangle$, and if the measurement reveals $|1\rangle$, the left/lower qubit is $X\,|\psi\rangle$. That is, we have the circuit:

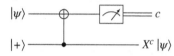

(3) We then consider the following circuit by adding the gates P^y and Z^d to the lower qubit, we obtain the output $Z^d P^y X^c\,|\psi\rangle$:[3]

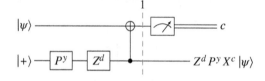

---

[3] We can verify the output as follows. Let $|\psi\rangle = \alpha\,|0\rangle + \beta\,|1\rangle$. Writing the lower qubit first, the circuit starts with:

$$(Z^d P^y \otimes I)\,|+\rangle\,|\psi\rangle = (Z^d P^y \otimes I)\frac{1}{\sqrt{2}}(|0\rangle + |1\rangle)(\alpha\,|0\rangle + \beta\,|1\rangle)$$

$$= (Z^d P^y \otimes I)\frac{1}{\sqrt{2}}(\alpha\,|00\rangle + \beta\,|01\rangle + \alpha\,|10\rangle + \beta\,|11\rangle)$$

$$= \frac{1}{\sqrt{2}}(\alpha\,|00\rangle + \beta\,|01\rangle + (-1)^d i^y \alpha\,|10\rangle + (-1)^d i^y \beta\,|11\rangle)$$

Then, applying CNOT, we have at step 1:

$$\frac{1}{\sqrt{2}}(\alpha\,|00\rangle + \beta\,|01\rangle + (-1)^d i^y \alpha\,|11\rangle + (-1)^d i^y \beta\,|10\rangle)$$

$$= \frac{1}{\sqrt{2}}((\alpha\,|0\rangle + (-1)^d i^y \beta\,|1\rangle)\,|0\rangle + ((-1)^d i^y \alpha\,|1\rangle + \beta\,|0\rangle)\,|1\rangle)$$

$$= \frac{1}{\sqrt{2}}(Z^d P^y(\alpha\,|0\rangle + \beta\,|1\rangle)\,|0\rangle + Z^d P^y(\alpha\,|1\rangle + \beta\,|0\rangle)\,|1\rangle)$$

$$= \frac{1}{\sqrt{2}}(Z^d P^y(\alpha\,|0\rangle + \beta\,|1\rangle)\,|0\rangle + Z^d P^y X(\alpha\,|0\rangle + \beta\,|1\rangle)\,|1\rangle)$$

$$= \frac{1}{\sqrt{2}}(Z^d P^y\,|\psi\rangle\,|0\rangle + Z^d P^y X\,|\psi\rangle\,|1\rangle)$$

So, if we measure the top qubit (the qubit on the right), we see that if the measurement reveals $|0\rangle$, the left/lower qubit is $Z^d P^y |\psi\rangle$, and if the measurement reveals $|1\rangle$, the left/lower qubit is $Z^d P^y X |\psi\rangle$.

(4) Next, we consider $X^a Z^b |\psi\rangle$ as the input to the circuit and apply R to it, so that we have the circuit:

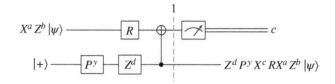

since the target for the CNOT in the circuit is $R X^a Z^b |\psi\rangle$.

(5) Next, we apply $P^{a \oplus y}$ to the lower qubit, so that we have the circuit:

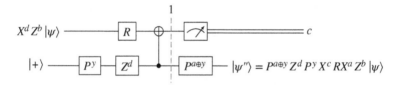

which is, effectively, the circuit in Figure 5.1. We just need to show that

$$|\psi''\rangle = P^{a \oplus y} Z^d P^y X^c R X^a Z^b |\psi\rangle = X^{a \oplus c} Z^{a \cdot (c \oplus y \oplus 1) \oplus b \oplus d \oplus y} R |\psi\rangle$$

via calculations (see **Aside**).

Aside: Show that $P^{a \oplus y} Z^d P^y X^c R X^a Z^b |\psi\rangle = X^{a \oplus c} Z^{a \cdot (c \oplus y \oplus 1) \oplus b \oplus d \oplus y} R |\psi\rangle$:

$P^{a \oplus y} Z^d P^y X^c R X^a Z^b |\psi\rangle \equiv P^{a \oplus y} Z^d P^y X^c X^a Z^a P^a R Z^b |\psi\rangle$

(ignoring the global phase, since $R X^a |\phi\rangle = (e^{i\pi/4})^a X^a Z^a P^a R |\varphi,\rangle$

for any $|\varphi\rangle$)

$= P^{a \oplus y} Z^d P^y X^c X^a Z^a Z^b P^a R |\psi\rangle$

(since $R Z^b = Z^b R$ and $P^a Z^b = Z^b P^a$)

$= Z^{a \cdot y} P^a P^y Z^d P^y X^c X^a Z^a Z^b P^a R |\psi\rangle$

(since $P^{a \oplus y} = P^{2a \cdot y + a + y} = Z^{a \cdot y} P^{a+y} = Z^{a \cdot y} P^a P^y$, $P^2 = Z$, and $P^4 = Z^2 = I$)

$= Z^{a \cdot y} P^a P^y P^y Z^d X^c X^a Z^a Z^b P^a R |\psi\rangle$

(since $Z^d P^y = P^y Z^d$)

$$= Z^{a \cdot y} Z^y Z^d P^a X^c X^a Z^a Z^b P^a R \ket{\psi}$$

(since $P^y P^y = Z^y, P^a Z^y = Z^y P^a$, and $P^a Z^d = Z^d P^a$)

$$= Z^{a \cdot y \oplus y \oplus d} P^a X^{a \oplus c} Z^{a \oplus b} P^a R \ket{\psi}$$

$$\equiv Z^{a \cdot y \oplus y \oplus d} Z^{a \cdot (a \oplus c)} X^{a \oplus c} P^a Z^{a \oplus b} P^a R \ket{\psi}$$

(since $P^a X^{a \oplus c} \equiv Z^{a \cdot (a \oplus c)} X^{a \oplus c} P^a$ by ignoring the global phase)

$$= Z^{a \cdot y \oplus y \oplus d} Z^{a \cdot (a \oplus c)} X^{a \oplus c} P^a P^a Z^{a \oplus b} R \ket{\psi}$$

(since $Z^{a \oplus b} P^a = P^a Z^{a \oplus b}$)

$$= Z^{a \cdot y \oplus y \oplus d} Z^{a \cdot (a \oplus c)} X^{a \oplus c} Z^a Z^{a \oplus b} R \ket{\psi}$$

(since $P^a P^a = P^{2a} = Z^a$)

$$= X^{a \oplus c} Z^{a \cdot y \oplus y \oplus d} Z^{a \cdot (a \oplus c)} Z^a Z^{a \oplus b} R \ket{\psi}$$

(since $Z^{a \cdot (a \oplus c)} X^{a \oplus c} = -X^{a \oplus c} Z^{a \cdot (a \oplus c)}$ and $Z^{a \cdot y \oplus y \oplus d} X^{a \oplus c} = -X^{a \oplus c} Z^{a \cdot y \oplus y \oplus d}$)

$$= X^{a \oplus c} Z^{a \cdot (c \oplus y \oplus 1) \oplus b \oplus d \oplus y} R \ket{\psi}$$

In Figure 5.1, if B sends $\ket{\psi''}$ to A, note that A has a, b, y, d, and c (also from B), and so, can obtain $R \ket{\psi}$ from $\ket{\psi''}$ as follows:

$$Z^{a \cdot (c \oplus y \oplus 1) \oplus b \oplus d \oplus y} X^{a \oplus c} \ket{\psi''} = R \ket{\psi}$$

We have seen how the client A can ask the server to perform quantum operations and decrypt the result from the server B. Table 5.1 summarizes the client's encrypted states and what is sent to the server, the operation performed on the server and computed state on the server (with the equivalent form showing the corresponding encryption keys); in short, the idea is

$$\ket{\psi} \xrightarrow{enc_{a,b}} enc_{a,b}(\ket{\psi}) \xrightarrow{U} U(enc_{a,b}(\ket{\psi})) = enc_{a',b'}(U \ket{\psi}) \xrightarrow{dec_{a',b'}} U \ket{\psi}$$

where we have enc using the quantum one-time pad, i.e. for randomly and uniformly chosen $a, b \in \{0, 1\}$, $enc_{a,b}(\ket{\psi}) = X^a Z^b \ket{\psi}$, and $dec_{a',b'} = Z^{b'} X^{a'}$. A core idea for this to work is to have the equivalence: $U (enc_{a,b}(\ket{\psi})) = enc_{a',b'}(U \ket{\psi}) = X^{a'} Z^{b'}(U \ket{\psi})$, with suitable values for a' and b', so that $dec_{a',b'}(enc_{a',b'}(U \ket{\psi})) = Z^{b'} X^{a'}(enc_{a',b'}(U \ket{\psi})) = Z^{b'} X^{a'} X^{a'} Z^{b'}(U \ket{\psi}) = U \ket{\psi}$.

But in general, a quantum circuit would use more than just one gate. A general quantum circuit can be decomposed into a sequence of gates, where the gates are from the set $\{X, Z, H, P, CNOT, R\}$, and hence, the reason for showing how each of these gates can be delegated to the server.

Assuming that qubits are prepared and held in a quantum register, the following protocol is used to delegate computations to a server:

Table 5.1 Table summarizing the protocol for computing with encrypted data for different gates for each qubit (single qubit gates) or each pair of qubits (for the *CNOT*).

From client	U on server	Computed on server	Keys for *dec*
$X^a Z^b \lvert \psi \rangle$	X	$X(X^a Z^b \lvert \psi \rangle) \equiv X^a Z^b(X \lvert \psi \rangle)$	$a' = a, b' = b$
$X^a Z^b \lvert \psi \rangle$	Z	$Z(X^a Z^b \lvert \psi \rangle) \equiv X^a Z^b(Z \lvert \psi \rangle)$	$a' = a, b' = b$
$X^a Z^b \lvert \psi \rangle$	H	$H(X^a Z^b \lvert \psi \rangle) \equiv X^b Z^a(H \lvert \psi \rangle)$	$a' = b, b' = a$
$X^a Z^b \lvert \psi \rangle$	P	$P(X^a Z^b \lvert \psi \rangle) \equiv X^a Z^{a \oplus b}(P \lvert \psi \rangle)$	$a' = a, b' = a \oplus b$
$X^a Z^b \otimes X^c Z^d \lvert \psi \rangle$	*CNOT*	$CNOT(X^a Z^b \otimes X^c Z^d \lvert \psi \rangle) \equiv$	$a' = a, b' = b \oplus d$
(2-qubit $\lvert \psi \rangle$)		$(X^a Z^{b \oplus d} \otimes X^{a \oplus c} Z^d)(CNOT \lvert \psi \rangle)$	$c' = a \oplus c, d' = d$
$X^a Z^b \lvert \psi \rangle$,	R (and server's	c and $X^{a''} Z^{b''}(R \lvert \psi \rangle)$	$a' = a'', b' = b''$
$x = a \oplus y,$	operations	where $a'' = a \oplus c$ and	
$Z^d P^y \lvert + \rangle$	in Figure 5.1)	$b'' = a \cdot (c \oplus y \oplus 1) \oplus b \oplus d \oplus y$	

a) We ignore global phases in the equivalences "$\equiv$." The keys for decrypting are (a', b') using $Z^{b'} X^{a'}$ (and (c', d') in the case of *CNOT*).

(1) The client encrypts its quantum register with the quantum one-time pad we saw earlier and then sends the contents of the register to the server (and the circuit to be executed).

(2) The server then performs the computations according to the circuit using the encrypted quantum data, according to the procedures shown in Table 5.1, with the client computing the encryption keys for the corresponding quantum wire after each operation.

(3) On completion, the server sends the output register to the client which then decrypts it using its computed keys.

The protocol in general is illustrated as follows, which implements a circuit $\mathcal{U} = U_1 \ldots U_n$.

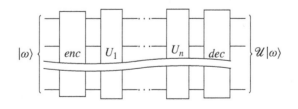

Effectively, we want to compute $\mathcal{U} \lvert \psi \rangle = dec(\mathcal{U}(enc(\lvert \psi \rangle)))$

Let us look at an example of the client delegating to the server the following (contrived) quantum circuit acting on three qubits denoted by $q_1 q_2 q_3$ (top is q_1 to bottom q_3) as shown below.

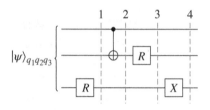

This circuit can be written as

$$(\mathbf{I} \otimes \mathbf{I} \otimes X)(\mathbf{I} \otimes R \otimes \mathbf{I})(CNOT_{q_1 q_2} \otimes \mathbf{I})(\mathbf{I} \otimes \mathbf{I} \otimes R)$$

where $CNOT_{q_1 q_2}$ denotes $CNOT$ on the top two qubits, the top qubit q_1 as control, that is, the client wants the server to help it compute:

$$|\psi'\rangle = (\mathbf{I} \otimes \mathbf{I} \otimes X)(\mathbf{I} \otimes R \otimes \mathbf{I})(CNOT_{q_1 q_2} \otimes \mathbf{I})(\mathbf{I} \otimes \mathbf{I} \otimes R)|\psi\rangle$$

First, with randomly selected $a, b, c, d, e, f \in \{0, 1\}$, the client encrypts the three qubit state $|\psi\rangle$ using the following:

$$(X^a Z^b \otimes X^c Z^d \otimes X^e Z^f)|\psi\rangle$$

which is sent to the server. There are two R gate operations. For the first R gate operation, the client chooses randomly $y_1, d_1 \in \{0, 1\}$ and sends $x_1 = a \oplus y_1$ and $Z^{d_1} p^{y_1} |+\rangle$ and the above encrypted state to the server. From Figure 5.1, the server computes c_1 sending this to the client which the client keeps, and at slice 1, the server has computed: $(X^a Z^b \otimes X^c Z^d \otimes X^{e \oplus c_1} Z^{e \cdot (c_1 \oplus y_1 \oplus 1) \oplus f \oplus d_1 \oplus y_1})(\mathbf{I} \otimes \mathbf{I} \otimes R)|\psi\rangle$.

Second, applying the CNOT to the top two qubits, at slice 2, the server has computed: $(X^a Z^{b \oplus d} \otimes X^{a \oplus c} Z^d \otimes X^{a''} Z^{b''})(CNOT_{q_1 q_2} \otimes \mathbf{I})(\mathbf{I} \otimes \mathbf{I} \otimes R)|\psi\rangle$, where $a'' = e \oplus c_1$ and $b'' = e \cdot (c_1 \oplus y_1 \oplus 1) \oplus f \oplus d_1 \oplus y_1$.

Third, for the second R gate operation, the client chooses randomly $y_2, d_2 \in \{0, 1\}$, and sends $x_2 = (a \oplus c) \oplus y_2$ and $Z^{d_2} p^{y_2} |+\rangle$ to the server (note that these could be sent earlier even before computations begin to avoid any wait times in the middle of computations). From Figure 5.1, the server computes c_2 sending this to the client which the client keeps, and at slice 3, the server has computed:

$$(X^a Z^{b \oplus d} \otimes X^{a'''} Z^{b'''} \otimes X^{a''} Z^{b''})(\mathbf{I} \otimes R \otimes \mathbf{I})(CNOT_{q_1 q_2} \otimes \mathbf{I})(\mathbf{I} \otimes \mathbf{I} \otimes R)|\psi\rangle,$$

where $a''' = (a \oplus c) \oplus c_2$, and $b''' = (a \oplus c) \cdot (c_2 \oplus y_2 \oplus 1) \oplus d \oplus d_2 \oplus y_2$.

Fourth and finally, applying the X gate to the qubit q_3 we have, at slice 4: $(X^a Z^{b \oplus d} \otimes X^{a'''} Z^{b'''} \otimes X^{a''} Z^{b''})(\mathbf{I} \otimes \mathbf{I} \otimes X)(\mathbf{I} \otimes R \otimes \mathbf{I})(CNOT_{q_1 q_2} \otimes \mathbf{I})$ $(\mathbf{I} \otimes \mathbf{I} \otimes R)|\psi\rangle = (X^a Z^{b \oplus d} \otimes X^{a'''} Z^{b'''} \otimes X^{a''} Z^{b''})|\psi'\rangle$ which the server then sends to the client as the result. To decrypt the above, the client can apply to the result the operation: $Z^{b \oplus d} X^a \otimes Z^{b'''} X^{a'''} \otimes Z^{b''} X^{a''}$, in order to get $|\psi'\rangle$! Note that the client has b, d, and a, and all the information to compute b''', a''', b'', and a'', which can be viewed as the keys to decrypt each of the qubits in the result, i.e. the pairs $(a, b \oplus d)$, (a''', b'''), and (a'', b''); the client effectively adjusts these keys according to the operations to compute the final keys.

An experimental demonstration of a protocol for delegated computation is explained in Fisher et al. [2014]. There is also an associated patent on quantum computing with encrypted data.[4]

5.2 How to Verify Delegated Private Quantum Computations

Note that in the above scheme for delegated computation, there is a question of trust, whether the client trusts the server to perform the computations that it wants the server to perform. While the server cannot know what input it is receiving from the client due to the use of the quantum one-time pad, the server might not be performing the operation required by the client! How can the client verify that when it wants the server to apply an X gate, for example, that the server actually performs an X gate operation? The server might actually do nothing or apply some other operation. Worse still, the server might not be a quantum computer after all and pretending to be one!

This is a broader issue of verifying quantum computations. Sometimes it is easy to verify the result from a quantum computation. For example, for the problem of factoring a large number using Shor's algorithm [Shor, 1994], once the client receives the answers from the server (which are suppose to be factors of a given large number), the client can do a multiplication to verify if the answers returned are indeed factors (and this would be true for other NP problems where verification can be done in polynomial time, say). But what about some problems where it is difficult, or intractable (at least if using only a classical computer or a client with limited quantum computing capabilities) to verify? This problem has been discussed extensively elsewhere – e.g. see the review in Gheorghiu et al. [2019].

Hence, the client, when delegating quantum computations to the server, doesn't just want to hide its data from the server (so-called *private* or *blind quantum computing* [Fitzsimons, 2017]) but also verify that the server is honest or performing computations correctly (i.e. *verifiability*). Here, we outline and discuss briefly the

---

4 See https://patentimages.storage.googleapis.com/7e/bf/2d/c77d16905722d8/US8897449.pdf [last accessed: 6/7/2022].

idea of using test runs to detect a dishonest server, from Broadbent [2018], i.e. the idea of test runs being interleaved with actual computation runs in such a way that the server does not know which type of run it is, so that the server cannot adapt or react to avoid or pass such tests.

Suppose the client wants to apply the circuit C to input $|0\rangle^{\otimes n}$. Note that without loss of generality, we can assume the input is $|0\rangle^{\otimes n}$ since the preparation for an input $|x\rangle$ can be integrated into the circuit, i.e. $C' |x\rangle = C|0\rangle^{\otimes n} = C'U|0\rangle^{\otimes n}$, where $C = C'U$ and $|x\rangle = U|0\rangle^{\otimes n}$.

The client randomly and uniformly chooses one of three possible types of runs (without telling the server its choice), sending to the server the quantum one-time pad encrypted input $|0\rangle^{\otimes n}$ and the circuit C; the client also sends to the server additional auxiliary qubits required for R-gate gadgets as we explain later:

- *Computation run*: The client delegates the actual required computation $C|0\rangle^{\otimes n}$ to the server
- *X-test run*: The client delegates the identity computation on $|0\rangle^{\otimes n}$ to the server
- *Z-test run*: The client delegates the identity computation on $|+\rangle^{\otimes n}$ to the server

The server does not know which run it has been delegated since it is also given the circuit C in any of the above type of run, and the server simply performs each operation in C. The idea is that the server's actions or operations is the same in any of the above runs, and it is what the client does on its side that determines which type of run has happened, as we explain below.

The protocol for the runs is similar to what we have seen in Section 5.1, as shown in Table 5.1, but there are key differences in the way the R-gate and the H gate are performed. Let's look at how the runs would proceed for each type of gate (these procedures which we simply call gadgets), which will then provide an idea of how the computation of an entire circuit comprising such gates might be delegated.

5.2.1 *X* Gate Gadget

Suppose the client wants to perform the X gate operation on an encrypted qubit $X^a Z^b |\psi\rangle$. We saw earlier that the server can perform an X gate operation on the encrypted qubit $X^a Z^b |\psi\rangle$ to get $XX^a Z^b |\psi\rangle \equiv X^a Z^b X |\psi\rangle$ which the client can decrypt using the keys a, b, i.e. using $Z^b X^a$ to get $X |\psi\rangle$. Another way to do this is actually have the server do nothing and just the client adjusting the keys to a', b', where $a' = a \oplus 1$ and $b' = b$; this works since if the server sends back the original input $X^a Z^b |\psi\rangle$, unchanged, then decrypting using a', b' gives:

$$Z^{b'} X^{a'} X^a Z^b |\psi\rangle = Z^b X^{a\oplus 1} X^a Z^b |\psi\rangle = Z^b X X^a X^a Z^b |\psi\rangle \equiv X |\psi\rangle$$

Note that the client adjusts its keys only in a computation run. Since it intends to compute the identity operation in the X-test and Z-test runs, the client does not adjust its keys when it wants to do a test run (while the server also does not need to do anything, like in the computation run).

5.2.2 *Z* Gate Gadget

We saw earlier how the client can get the server to do a Z gate on an encrypted qubit. Similar to the X gate, the server can do nothing (or send back the original encrypted qubit unchanged) and the client just adjusts its keys, this time to $a' = a$ and $b' = b \oplus 1$ so that we have:

$$Z^{b'} X^{a'} X^a Z^b \, |\psi\rangle = Z^{b \oplus 1} X^a X^a Z^b \, |\psi\rangle = ZZ^b Z^b \, |\psi\rangle = Z \, |\psi\rangle$$

As in the case of the X-gate, the client adjusts its keys only in a computation run. Since it intends to compute the identity operation in the X-test and Z-test runs, the client does not adjust its keys (while the server also does not need to do anything, like in the computation run).

5.2.3 *CNOT* Gate Gadget

In the case of the client wanting the server to perform the $CNOT$ gate, the approach for the computation run is as in Table 5.1: applying a CNOT gate on qubits i and j, where i is the control and j is the target, suppose qubit i has key (a_i, b_i) and qubit j has key (a_j, b_j), then, the server performs the CNOT between these two qubits, and the client adjusts the keys as follows to have the new keys (a_i', b_i') and (a_j', b_j'): $a_i' = a_i, b_i' = b_i \oplus b_j$, and $a_j' = a_i \oplus a_j, b_j' = b_j$. In either a X-test and Z-test run, the server still does the CNOT, but the client does not adjust its keys since $CNOT(|0\rangle \, |0\rangle) = |0\rangle \, |0\rangle$ and $CNOT(|+\rangle \, |+\rangle) = |+\rangle \, |+\rangle$.

5.2.4 *R* Gate Gadget

Let us look at the way the R operation is carried out, i.e. the R gate "gadget," as given in Figure 5.2.

Compared to the previous R gadget in Figure 5.1, here, the R gate is "done" on the client side. In fact, similar to Section 5.1, for this gadget, the client does not need to be able to actually perform these gate operations on arbitrary quantum states, but just be able to *prepare* a state chosen randomly and uniformly from one of the following states, here specified in terms of results of operations: $\{R \, |+\rangle, R \, |-\rangle, PR \, |+\rangle, PR \, |-\rangle\}$ (similar to Assumption 1 earlier); this is equivalent to the client choosing randomly and uniformly $d, e, y \in \{0, 1\}$ and computing $X^d Z^e P^y R \, |+\rangle$ since:

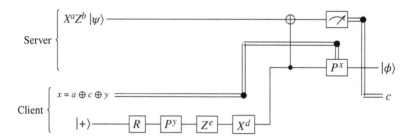

Figure 5.2 A slightly modified protocol for an R-gate for a computation run, where the output from the server is $|\phi\rangle = X^{a\oplus c}Z^{(a\oplus c)\cdot(d\oplus y)\oplus a\oplus b\oplus c\oplus e\oplus y}R\,|\psi\rangle$. The client chooses randomly and uniformly $d, e, y \in \{0, 1\}$. This can be called an R-gate "gadget."

d	e	y	$X^d Z^e P^y R\,|+\rangle$
0	0	0	$R\,|+\rangle$
0	0	1	$PR\,|+\rangle$
0	1	0	$ZR\,|+\rangle = RZ\,|+\rangle = R\,|-\rangle$
0	1	1	$ZPR\,|+\rangle = PZR\,|+\rangle = PRZ\,|+\rangle = PR\,|-\rangle$
1	0	0	$XR\,|+\rangle \equiv ZPRX\,|+\rangle = ZPR\,|+\rangle = PZR\,|+\rangle = PRZ\,|+\rangle = PR\,|-\rangle$
1	0	1	$XPR\,|+\rangle \equiv ZPXR\,|+\rangle \equiv ZP\,ZPRX\,|+\rangle = ZPZPR\,|+\rangle = R\,|-\rangle$
1	1	0	$XZR\,|+\rangle = -ZXR\,|+\rangle \equiv -ZZPRX\,|+\rangle = -PR\,|+\rangle \equiv PR\,|+\rangle$
1	1	1	$XZPR\,|+\rangle \equiv RX\,|+\rangle = R\,|+\rangle$

where in the above, we are using the identities: $XZ = -ZX$, $PZ = ZP$, $ZR = RZ$, $RX \equiv XZPR$, $XR \equiv ZPRX$, $PR = RP$, $XP \equiv ZPX$, $X\,|+\rangle = |+\rangle$, $Z\,|+\rangle = |-\rangle$, and $P^2 = Z$, where the equivalences are due to ignoring global phases.

The circuit in Figure 5.2 can be derived as follows, similar to how we derived the R-gate gadget in Section 5.1.

1) Let's start with the following circuit, similar to the one in step (3) of the previous derivation for Figure 5.1, but with the additional R gate before the P^y gate in the bottom wire and input being $X^a Z^b\,|\psi\rangle$, so that the output becomes $Z^e P^y R X^c X^a Z^b\,|\psi\rangle$:

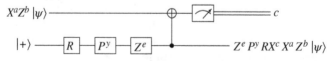

One can check as we have done for step (3) in the previous derivation that the output is $Z^e P^y RX^c X^a Z^b |\psi\rangle$.

2) Then, we add the X^d gate after the control in the bottom wire, so that the output becomes $X^d Z^e P^y RX^c X^a Z^b |\psi\rangle$:

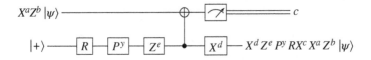

3) Then, we add the gate $P^x = P^{a \oplus c \oplus y}$ after the X^d gate to the bottom wire as follows, so that the output becomes $|\varphi\rangle = P^{a \oplus c \oplus y} X^d Z^e P^y RX^c X^a Z^b |\psi\rangle$:

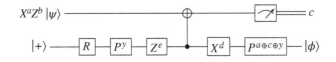

Now, we rephrase the output:

$$P^x X^d Z^e P^y RX^c X^a Z^b |\psi\rangle = P^{a \oplus c \oplus y} X^d Z^e P^y RX^{a \oplus c} Z^b |\psi\rangle$$

(since $x = a \oplus c \oplus y$)

$$\equiv P^{a \oplus c \oplus y} X^d Z^e P^y X^{a \oplus c} Z^{a \oplus c} P^{a \oplus c} RZ^b |\psi\rangle$$

(since $RX^{a \oplus c} \equiv X^{a \oplus c} Z^{a \oplus c} P^{a \oplus c} R$, ignoring the global phase)

$$= P^{a \oplus c \oplus y} X^d Z^e P^y X^{a \oplus c} P^{a \oplus c} Z^{a \oplus c} Z^b R |\psi\rangle$$

(using $PZ = ZP, ZR = RZ$)

$$\equiv P^{a \oplus c \oplus y} X^d Z^e P^y Z^{a \oplus c} P^{a \oplus c} X^{a \oplus c} Z^{a \oplus c} Z^b R |\psi\rangle$$

(using $XP \equiv ZPX$, and ignoring the global phase)

$$= -P^{a \oplus c \oplus y} X^d Z^e P^y P^{a \oplus c} X^{a \oplus c} Z^b R |\psi\rangle$$

(using $PZ = ZP, -ZX = XZ$ and $Z^{a \oplus c} Z^{a \oplus c} = I$)

$$= -P^{a \oplus c \oplus y} X^d P^y Z^e P^{a \oplus c} X^{a \oplus c} Z^b R |\psi\rangle$$

(using $PZ = ZP$)

$$\equiv -P^{a \oplus c \oplus y} Z^{d \cdot y} P^y X^d Z^e P^{a \oplus c} X^{a \oplus c} Z^b R |\psi\rangle$$

(using $X^d P^y \equiv Z^{d \cdot y} P^y X^d$, ignoring the global phase)

$$= P^{a \oplus c \oplus y} Z^{d \cdot y} P^y Z^e X^d P^{a \oplus c} X^{a \oplus c} Z^b R |\psi\rangle$$

(using $XZ = -ZX$)

$$\equiv P^{a \oplus c \oplus y} Z^{d \cdot y} P^y Z^e Z^{d \cdot (a \oplus c)} P^{a \oplus c} X^d X^{a \oplus c} Z^b R |\psi\rangle$$

(using $X^d P^{a \oplus c} \equiv Z^{d \cdot (a \oplus c)} P^{a \oplus c} X^d$, ignoring the global phase)

$$= P^{(a\oplus c)\oplus y}P^y P^{a\oplus c}Z^{d\cdot y\oplus e\oplus d\cdot(a\oplus c)}X^d X^{a\oplus c}Z^b R\,|\psi\rangle$$

(using $PZ = ZP$)

$$= Z^{y\cdot(a\oplus c)}P^{(a\oplus c)}P^y P^y P^{a\oplus c}Z^{d\cdot y\oplus e\oplus d\cdot(a\oplus c)}X^d X^{a\oplus c}Z^b R\,|\psi\rangle$$

(using $Z^{y\cdot(a\oplus c)}P^{(a\oplus c)}P^y = P^{(a\oplus c)\oplus y}$)

$$= X^d X^{a\oplus c}Z^{y\cdot(a\oplus c)}Z^{a\oplus c\oplus y}Z^{d\cdot y\oplus e\oplus d\cdot(a\oplus c)}Z^b R\,|\psi\rangle$$

(using $P^y P^y = Z^y$, $P^{a\oplus c}P^{a\oplus c} = Z^{a\oplus c}$ and $ZX = -XZ$)

$$= X^{a\oplus c\oplus d}Z^{y\cdot(a\oplus c)\oplus a\oplus c\oplus y\oplus d\cdot y\oplus e\oplus d\cdot(a\oplus c)\oplus b}R\,|\psi\rangle$$

$$= X^{a\oplus c\oplus d}Z^{y\cdot(a\oplus c)\oplus a\oplus c\oplus y\oplus d\cdot y\oplus e\oplus d\cdot(a\oplus c)\oplus b\oplus d\cdot d\oplus d}R\,|\psi\rangle$$

(since $d\cdot d\oplus d = d\oplus d = 0$)

$$= X^{a\oplus c\oplus d}Z^{(a\oplus c\oplus d)\cdot(d\oplus y)\oplus a\oplus b\oplus c\oplus d\oplus e\oplus y}R\,|\psi\rangle$$

$$= X^{a\oplus c'}Z^{(a\oplus c')\cdot(d\oplus y)\oplus a\oplus b\oplus c'\oplus e\oplus y}R\,|\psi\rangle$$

(where we let $c' = c\oplus d$)

4) Then, we make use of the following equivalence about the *CNOT* and X gates:

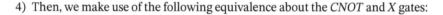

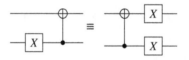

so that if we add an X-gate after the target to the top wire, we have:

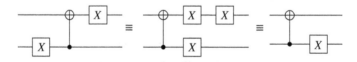

And so, the circuit from step (3) can be equivalently drawn as follows:

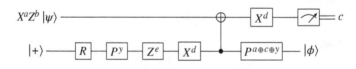

where $|\phi\rangle = X^{a\oplus c'}Z^{(a\oplus c')\cdot(d\oplus y)\oplus a\oplus b\oplus c'\oplus e\oplus y}R\,|\psi\rangle$ with $c' = c\oplus d$, from the calculations above.

5) Finally, we observe that for any qubit $\alpha\,|0\rangle + \beta\,|1\rangle$ emerging after the target in the top wire, since $X^d(\alpha\,|0\rangle + \beta\,|1\rangle) = \alpha\,|0\oplus d\rangle + \beta\,|1\oplus d\rangle$, applying X^d to the qubit and then measuring will return the result c (where $c = 0\oplus d$ or $c = 1\oplus d$), but if $c' = c\oplus d$, then the equivalent circuit is as follows:

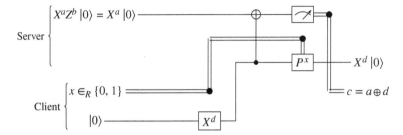

Figure 5.3 The R-gate gadget in an X-test run. Note that the client randomly and uniformly chooses $x, d \in \{0, 1\}$, the server does the same operations as in a computation run (compare with Figure 5.2), and note that operations on the client is such that the identity operation is carried out (the identity operation is *up to encryption*, i.e. as long as the client keeps track of the key, in this case, after this gadget, the key has changed from a to d; the client can apply X^d to get back $|0\rangle$). Also, there is a test – after receiving c from the server, the client verifies that $c = a \oplus d$.

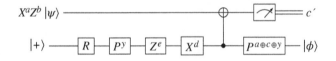

which is the circuit in Figure 5.2.

In the X-test run and the Z-test run, different R-gate gadgets are used.

Figure 5.3 shows the R gate used in the X-test run. Note that the gadget has a test to check that the server's output c satisfies $c = a \oplus d$. This is because the state after the target and before measurement in the top wire is $X^{a\oplus d} |0\rangle$, which upon measurement should provide $c = a \oplus d$ if the server has been doing the right thing, and noting that the output from the bottom wire would be $P^x X^d |0\rangle \equiv X^d |0\rangle$ (ignoring the global phase).

Figure 5.4 shows the R gate used in the Z-test run. The output $X^c Z^{b\oplus d\oplus y} |+\rangle$ is obtained as shown via calculations. From the circuit, (similar to the idea in step (3) of the derivation for Figure 5.1) the output is

$$P^x Z^d P^y X^c Z^b |+\rangle = P^y P^y Z^d X^c Z^b |+\rangle = Z^y Z^d X^c Z^b |+\rangle = X^c Z^{b\oplus d\oplus y} |+\rangle$$

(since $x = y$, $ZP = PZ$, and $P^y P^y = Z^y$)

5.2.5 H Gate Gadget

The H gate is what makes the Z-test run needed since it changes $|0\rangle$ to $|+\rangle$. There are some difficulties here if the H gate is simply applied by the server like we did in Section 5.1, see Table 5.1 since, this time, we don't have just computation

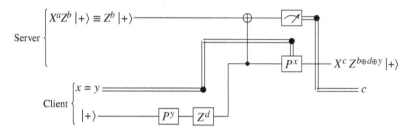

Figure 5.4 The *R*-gate gadget in a *Z*-test run. Note that the client randomly and uniformly chooses $d, y \in \{0, 1\}$, the server does the same operations as in a computation run and the *X*-test run (compare with Figures 5.2 and 5.3), and note that operations on the client is such that the identity operation is carried out (the identity operation is *up to encryption*, i.e. as long as the client keeps track of the key, in this case, after this gadget, the key has changed from b to $(c, b \oplus d \oplus y)$).

runs but also test runs. For example, suppose each qubit i is in a state of the form $X^{a_i} Z^{b_i} |0\rangle$, for some a_i and b_i, and $i \in \{1, \ldots n\}$, and a H gate is applied to one of them (say j) to become $H(X^{a_j} Z^{b_j} |0\rangle) = X^{b_j} Z^{a_j} H |0\rangle = X^{b_j} Z^{a_j} |+\rangle$ (from Table 5.1); hence, this would change the state of a qubit to $|+\rangle$ while the other qubits remain $|0\rangle$, up to encryption. This complicates matters if later a *CNOT* gate is applied between this qubit j and some other qubit $i \neq j$, that is, we would have to deal with $CNOT_{ij}(|0\rangle |+\rangle)$, since we no longer will be doing $CNOT(|0\rangle |0\rangle)$ (in the *X*-test run) and correspondingly, we no longer will be doing $CNOT(|+\rangle |+\rangle)$ (in the *Z*-test run) so that the *CNOT* gate gadget we used earlier might not work.

The solution given by Broadbent [2018] is to perform the following operations (which we call U_H) for each H gate: $U_H := HPHPHPH = HRRHRRHRRH$, since $P = RR$; each H gate in the circuit is replaced by the U_H operation, in any type of run.

In a computation run, due to the equivalence: $HPHPHPH = H$, the replacements do not change the circuit's outcome, and the effect of applying U_H is equivalent to applying H.

In a test run, for each R gate operation in U_H, we use the corresponding R gate gadget (Figure 5.3 for the *X*-test run and Figure 5.4 for the *Z*-test run), which computes the identity operation (up to encryption), and so we have:

$$U_H = HPHPHPH = HRRHRRHRRH = HHHH$$

and since $HHHH = \mathbf{I}$, effectively, we have the identity operation. Note that in an *X*-test run, after applying the first H in U_H to a qubit j, if initially the qubit is in a state of the form $X^{a_j} Z^{b_j} |0\rangle$, it goes into a state of the form $X^{a'_j} Z^{b'_j} |+\rangle$ for some appropriate a'_j, b'_j, i.e. the applicable R gate gadget for the next RR operation is now that for the *Z*-test run in Figure 5.4 (that is, effectively, the *X*-test run changes to a *Z*-test run). And this R gate gadget is used twice (for the first RR). Then, the second H gate is applied, and the state of qubit j changes back to one in the form $X^{a''_j} Z^{b''_j} |0\rangle$ for some appropriate a''_j, b''_j, so that the applicable R-gate gadget is the one for the X-test run in Figure 5.3 for the second RR. Thereafter, the third H is applied, and similarly, the qubit state changes over to a form where the applicable R gate gadget for the next RR operation is now that for the Z-test run in Figure 5.4. Finally, the last H is applied and the qubit reverts back to a form equivalent (up to encryption) to $|0\rangle$. In a Z-test run, the sequence of applicable R gate gadgets is then the one in Figure 5.3 for the first RR, then Figure 5.4 for the second RR and Figure 5.3 for the final RR.

So, the server does the same computations (on its side) in any type of run, which is its side of U_H, but the client decides whether this computation is to apply H (i.e. $U_H = H$) or the identity operation (i.e. $U_H = \mathbf{I}$) depending on the R gate gadget used (that is, on the type of run the client has chosen).

Once a circuit is delegated to the server, in a computation run, the server and client proceed accordingly as above using the appropriate gadget according to the gates in the circuit. The client adjusts and computes the keys so that it can decrypt the final results returned by the server.

In a test run, verifications can be carried out (e.g. for each R gate as in Figure 5.3), and the client can abort when some verification fails. Also, the server can be instructed to measure the outcome and return to the client the result; for example, the client can check that the result should correspond to $|0\rangle^{\otimes n}$ in a X-test run (given that an identity operation should have been carried out), to detect if the server has behaved. Wrong results after a number of test runs or failed verifications might help the client conclude there is something wrong or unusual about the server, but if the server consistently showed correct results in test runs, it can then gain the client's trust. However, the approach is probabilistic; unluckily for the client, the server might happen to just behave during test runs and misbehave during computation runs!

We see that for any circuit the client delegates to it using the above protocol, the server does the same operations, regardless of which type of run it is. Hence, the server does not know if it is actually performing a computation run or being tested, and so, the idea is that the server needs to perform correctly in any run, or risk being caught out. Note that the client needs to be able to *prepare* certain states, however, for this protocol to work. The original description of the above procedure in Broadbent [2018] considers the procedure involving the client side and server side algorithms from a computational complexity perspective, as a quantum-prover interactive proof (QPIP) system for decision problems, with the server as the *prover* and the client as *verifier*. We won't go into the details of this here but note that if the client provides to the server a circuit that has certain properties, an honest prover should be able to demonstrate results to the client consistent with those properties, with client and server running the algorithms as above.

A review of protocols for verifying delegated quantum computations is by Gheorghiu et al. [2019], including techniques based on measurement-based quantum computation (which we did not cover in this book). An implementation of verifiable blind quantum computing is described in Drmota et al. [2023], where the server is a trapped-ion quantum computer and the client is a photonic detection system, with the client and server networked using a fiber-optic quantum link.

5.3 Quantum Computing-as-a-Service

When time on quantum computers is offered as a resource, or as a service, via the Internet by a cloud service provider, we have the notion of Quantum Computing-as-a-Service (or QaaS). This is already being done by major cloud service providers mentioned in Chapter 1. More generally, the execution of quantum circuits would be embedded into a program containing both classical computations and quantum computations – a programming model is required for this, e.g. a Python program with library calls to execute (predefined or user defined) quantum circuits. Response times (as well as queuing times), reliability, availability, capacity (e.g. the number and "quality" of qubits supported), quality-of-service guarantees, and economic cost of quantum program (circuit) executions are among factors to be considered when choosing one or more quantum computer providers [Moguel et al., 2022; Sodhi and Kapur, 2021; Ravi et al., 2021]. Apart from high-level abstractions, a Web front end can be used to help users work with programs that interact with quantum computers in the backend [Grossi et al., 2021]. On the cloud server end, when the same quantum computer (supporting some fixed number of qubits) is being used for running multiple quantum programs (circuits), strategies to do this mapping to maintain efficient execution is required, e.g. see Liu and Dou [2021]. Also, scenarios of multiple organizations (each with its quantum computer platform) sharing quantum computing resources is discussed in Ngoenriang et al. [2022] – distributed quantum computing becomes important here to allow the quantum computers from multiple organizations to be pooled together for large-scale quantum computations.

5.4 Summary

We have reviewed mechanisms for a client, with limited quantum capabilities, to delegate quantum computations to a more powerful quantum computer. We looked at how delegated quantum computations can be obscured so that the server cannot know the client's inputs while still being able to perform the computations for the client. We also saw how the client can verify the server's computations, or at least detect, with some probability, if the server did not perform the delegated computations correctly. We also briefly discussed the notion of QaaS.

Studies have been done on how a resource-limited client can trust the results from (what it thinks is a) quantum computer, e.g. by cross verification using multiple machines [Greganti et al., 2021]. Mahadev [2022] addressed the question of whether a computer is indeed a quantum one and whether it can be verified by a classical device.

References

Anne Broadbent. Delegating private quantum computations. *Canadian Journal of Physics*, 93(9):941–946, 2015.

Anne Broadbent. How to verify a quantum computation. *Theory of Computing*, 14(11):1–37, 2018. doi: 10.4086/toc.2018.v014a011. URL https://theoryofcomputing.org/articles/v014a011.

Andrew M. Childs. Secure assisted quantum computation. *Quantum Information and Computation*, 5(6):456–466, Sep 2005. ISSN 1533-7146.

P. Drmota, D. P. Nadlinger, D. Main, B. C. Nichol, E. M. Ainley, D. Leichtle, A. Mantri, E. Kashefi, R. Srinivas, G. Araneda, C. J. Ballance, and D. M. Lucas. Verifiable blind quantum computing with trapped ions and single photons. CoRR, May 2023. URL https://doi.org/10.48550/arXiv.2305.02936.

K. A. G. Fisher, A. Broadbent, L. K. Shalm, Z. Yan, J. Lavoie, R. Prevedel, T. Jennewein, and K. J. Resch. Quantum computing on encrypted data. *Nature Communications*, 5(1):3074, 2014.

Joseph F. Fitzsimons. Private quantum computation: an introduction to blind quantum computing and related protocols. *npj Quantum Information*, 3(1):23, 2017.

Alexandru Gheorghiu, Theodoros Kapourniotis, and Elham Kashefi. Verification of quantum computation: an overview of existing approaches. *Theory of Computing Systems*, 63(4):715–808, 2019.

C. Greganti, T. F. Demarie, M. Ringbauer, J. A. Jones, V. Saggio, I. Alonso Calafell, L. A. Rozema, A. Erhard, M. Meth, L. Postler, R. Stricker, P. Schindler, R. Blatt, T. Monz, P. Walther, and J. F. Fitzsimons. Cross-verification of independent quantum devices. *Physical Review X*, 11:031049, Sep 2021. doi: 10.1103/PhysRevX.11.031049.

Michele Grossi, Luca Crippa, Antonello Aita, Giacomo Bartoli, Vito Sammarco, Eleonora Picca, N. Said, Filippo Tramonto, and Federico Mattei. A serverless cloud integration for quantum computing. CoRR, abs/2107.02007, 2021. URL https://arxiv.org/abs/2107.02007.

Lei Liu and Xinglei Dou. QuCloud: a new qubit mapping mechanism for multi-programming quantum computing in cloud environment. In *2021 IEEE International Symposium on High-Performance Computer Architecture (HPCA)*, pages 167–178, 2021. doi: 10.1109/HPCA51647.2021.00024.

Urmila Mahadev. Classical verification of quantum computations. *SIAM Journal on Computing*, 51(4):1172–1229, 2022.

Enrique Moguel, Javier Rojo, David Valencia, Javier Berrocal, Jose Garcia-Alonso, and Juan M. Murillo. Quantum service-oriented computing: current landscape and challenges. *Software Quality Journal*, 30:983–1002, 2022.

Napat Ngoenriang, Minrui Xu, Jiawen Kang, Dusit Niyato, Han Yu, and Xuemin Shen. DQC^{2}O: distributed quantum computing for collaborative optimization in future networks. CoRR, abs/2210.02887, 2022. doi: 10.48550/arXiv.2210.02887. URL https://doi.org/10.48550/arXiv.2210.02887.

Gokul Subramanian Ravi, Kaitlin N. Smith, Prakash Murali, and Frederic T. Chong. Adaptive job and resource management for the growing quantum cloud. In *2021 IEEE International Conference on Quantum Computing and Engineering (QCE)*, pages 301–312, 2021. doi: 10.1109/QCE52317.2021.00047.

P. W. Shor. Algorithms for quantum computation: discrete logarithms and factoring. In *Proceedings 35th Annual Symposium on Foundations of Computer Science*, pages 124–134, 1994. doi: 10.1109/SFCS.1994.365700.

Balwinder Sodhi and Ritu Kapur. Quantum computing platforms: assessing the impact on quality attributes and SDLC activities. In *2021 IEEE 18th International Conference on Software Architecture (ICSA)*, pages 80–91, 2021. doi: 10.1109/ICSA51549.2021.00016.

6

The Quantum Internet

This chapter discusses some of the key concepts and ideas of the quantum Internet, including the central role that entanglement plays, the concepts of entanglement swapping and entanglement purification, and briefly reviews the notion of quantum repeaters and quantum Internet architectures. We will also look at quantum computations in relation to the quantum Internet.

6.1 Entanglement Over Longer Distances

So far we have seen that entanglement is key to many distributed quantum computing protocols and computations. Any two nodes which want to do computations together (e.g. perform a distributed – CNOT operation) need to share entangled qubits, even when the two nodes are geographically far apart. One can think of the quantum Internet as a "black box" from which nodes can request an entanglement between two qubits (on two nodes), or roughly speaking, providing "entanglement-as-a-service" (with the associated service-based ideas of "quality" of the entangled pair and "efficiency," e.g. generating entangled qubits in a timely manner). How is such an entangled pair generated? Answering this question in detail will take us further into the physics which we will not do so here. But generally, there are a number of mechanisms to generate an entangled pair (e.g. a Bell state) between two nodes.[1] One key point, however, is that due to decoherence affecting entangled qubits, distance is a limitation – e.g. it is hard to directly transmit a qubit over long distances without it suffering loss – so that generating an entangled pair between two nodes far apart over long distances requires what has been called *quantum repeaters* (analogous to repeaters for classical networks, but a totally different technology), which has a mechanism to take two entangled pairs (e.g. between qubits A on node 1 and A' on node 2, and

---

1 A readable overview is at https://www.forbes.com/sites/chadorzel/2017/02/28/how-do-you-create-quantum-entanglement/?sh=2664a2c11732 [last accessed: 15/11/2022].

From Distributed Quantum Computing to Quantum Internet Computing: An Introduction, First Edition. Seng W. Loke.

between qubits B' on node 2 and B on node 3) and generate an entangled pair over a longer distance (e.g. between A on node 1 and B on node 3). Key to this mechanism is the notion of *entanglement swapping*, which we will look into. We will also look into a mechanism to transmit quantum information robustly over longer distances using *tree-cluster states*. But before all that, we need the idea of Bell State Measurement.

6.1.1 Bell States and Bell State Measurement

The four Bell states are $|\Phi^+\rangle = (|00\rangle + |11\rangle)/\sqrt{2}$, $|\Phi^-\rangle = (|00\rangle - |11\rangle)/\sqrt{2}$, $|\Psi^+\rangle = (|01\rangle + |10\rangle)/\sqrt{2}$, and $|\Psi^-\rangle = (|01\rangle - |10\rangle)/\sqrt{2}$. Since they are orthogonal, they form a basis, the *Bell basis*. For example, a state $|\psi\rangle$ written in the basis $\{|00\rangle, |01\rangle, |10\rangle, |11\rangle\}$:

$$|\psi\rangle = \alpha|00\rangle + \beta|11\rangle + \gamma|01\rangle + \delta|10\rangle$$

where $|\alpha|^2 + |\beta|^2 + |\gamma|^2 + |\delta|^2 = 1$, can be rewritten in the Bell basis as

$$|\psi\rangle = \frac{\alpha + \beta}{\sqrt{2}}|\Phi^+\rangle + \frac{\alpha - \beta}{\sqrt{2}}|\Phi^-\rangle + \frac{\gamma + \delta}{\sqrt{2}}|\Psi^+\rangle + \frac{\gamma - \delta}{\sqrt{2}}|\Psi^-\rangle$$

A Bell state measurement asks which of the Bell states $|\psi\rangle$ is in.

How can we measure in the Bell basis if we can only measure in the Z basis (due to the instruments available, for example)? See **Aside**.

Aside: Note that measuring a qubit in the X basis is applying the Hadamard gate first and then measuring in the Z basis in the sense that prob(+1) (or the probability of getting the state $|0\rangle$) in this case is the same as prob(+1) when measuring in the X basis.[2] The Pauli X measurement $|\psi\rangle \!-\!\boxed{\nearrow}^{X}$ can be implemented as $|\psi\rangle \!-\!\boxed{H}\!-\!\boxed{\nearrow}^{Z}$ since prob(+1) are the same in both.

Similarly, suppose $|\psi\rangle$ is a two-qubit state, and if we want to measure in the Bell basis but only can measure in the basis $\{|00\rangle, |01\rangle, |10\rangle, |11\rangle\}$, we need a U such that: $U|\Phi^+\rangle = |00\rangle$, $U|\Phi^-\rangle = |10\rangle$, $U|\Psi^+\rangle = |01\rangle$, and $U|\Psi^-\rangle = |11\rangle$, and use it as follows (where the $|\psi\rangle$ below is a two-qubit state):

---

2 See https://www.youtube.com/watch?v=7jMR-ey-cJc for more details on this [last accessed: 15/112022].

What is U? It is a CNOT gate followed by a Hadamard gate on the first qubit:

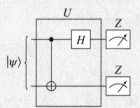

The circuit above enables us to measure the two-qubit state in the *Bell basis*, i.e. to do a Bell state measurement. (In general, measurement in any basis can be implemented by a unitary gate followed by measurement of each qubit in the Z basis.) Note that in the teleportation protocol in Chapter 2, we actually performed a Bell state measurement with the first two qubits at A (the qubit to be teleported and one of the qubits in the entangled pair).

6.1.2 Entanglement Swapping

From the previous definitions of Bell states, we have the Bell pair identities[3] :

$$|00\rangle = (|\Phi^+\rangle + |\Phi^-\rangle)/\sqrt{2} \tag{6.1}$$

$$|01\rangle = (|\Psi^+\rangle + |\Psi^-\rangle)/\sqrt{2} \tag{6.2}$$

$$|10\rangle = (|\Psi^+\rangle - |\Psi^-\rangle)/\sqrt{2} \tag{6.3}$$

$$|11\rangle = (|\Phi^+\rangle - |\Phi^-\rangle)/\sqrt{2} \tag{6.4}$$

Writing out the initial state of two entangled pairs, A with A' sharing $|\Phi^+\rangle_{AA'}$ (adding subscripts for clarity) and B' with B sharing $|\Phi^+\rangle_{B'B}$, as illustrated in Figure 6.1.

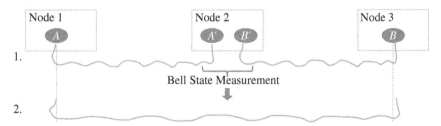

Figure 6.1 Illustration of entanglement swapping, from (1) entangled pairs between A at node 1 and A' at node 2 and between B' at node 2 and B at node 3 to (2) an entangled pair between A and B, after the Bell State Measurement at node 2.

---

3 See also https://www.youtube.com/watch?v=1zVvoADzXs4 [last accessed: 15/11/2022].

$$|\psi\rangle_{AA'B'B} = |\Phi^+\rangle_{AA'} \otimes |\Phi^+\rangle_{B'B}$$

$$= (|00\rangle + |11\rangle)/\sqrt{2} \otimes (|00\rangle + |11\rangle)/\sqrt{2}$$

(substituting using (6.1) twice)

$$= \frac{1}{2}(|0000\rangle + |0011\rangle + |1100\rangle + |1111\rangle)$$

$$= \frac{1}{2}(|0\rangle\,|00\rangle\,|0\rangle + |0\rangle\,|01\rangle\,|1\rangle + |1\rangle\,|10\rangle\,|0\rangle + |1\rangle\,|11\rangle\,|1\rangle)$$

$$= \frac{1}{2\sqrt{2}}(|0\rangle_A(|\Phi^+\rangle + |\Phi^-\rangle)_{A'B'}|0\rangle_B + |0\rangle_A(|\Psi^+\rangle + |\Psi^-\rangle)_{A'B'}|1\rangle_B$$

$$+ |1\rangle_A(|\Psi^+\rangle - |\Psi^-\rangle)_{A'B'}|0\rangle_B + |1\rangle_A(|\Phi^+\rangle - |\Phi^-\rangle)_{A'B'}|1\rangle_B)$$

(from substituting using (6.1)–(6.4) and reintroducing subscripts)

$$= \frac{1}{2\sqrt{2}}((|\Phi^+\rangle + |\Phi^-\rangle)_{A'B'}|00\rangle_{AB} + (|\Psi^+\rangle + |\Psi^-\rangle)_{A'B'}|01\rangle_{AB}$$

$$+ (|\Psi^+\rangle - |\Psi^-\rangle)_{A'B'}|10\rangle_{AB} + (|\Phi^+\rangle - |\Phi^-\rangle)_{A'B'}|11\rangle_{AB})$$

(by reordering qubits from $AA'B'B$ to $A'B'AB$)

$$= \frac{1}{2\sqrt{2}}|\Phi^+\rangle_{A'B'}|00\rangle_{AB} + \frac{1}{2\sqrt{2}}|\Phi^+\rangle_{A'B'}|11\rangle_{AB}$$

$$+ \frac{1}{2\sqrt{2}}|\Psi^+\rangle_{A'B'}|01\rangle_{AB} + \frac{1}{2\sqrt{2}}|\Psi^+\rangle_{A'B'}|10\rangle_{AB}$$

$$+ \frac{1}{2\sqrt{2}}|\Psi^-\rangle_{A'B'}|01\rangle_{AB} - \frac{1}{2\sqrt{2}}|\Psi^-\rangle_{A'B'}|10\rangle_{AB}$$

$$+ \frac{1}{2\sqrt{2}}|\Phi^-\rangle_{A'B'}|00\rangle_{AB} - \frac{1}{2\sqrt{2}}|\Phi^-\rangle_{A'B'}|11\rangle_{AB}$$

(rewriting according to Bell states in $A'B'$)

After Bell state measurement of $A'B'$, the central node (node 2) needs to tell nodes 1 and 3 that the measurement has completed, and nodes 1 and 3 then shares an entangled state (one of the Bell states, with equal probability). For example, if measurement of qubits $A'B'$ resulted in the state $|\Phi^+\rangle$, which occurs with probability $|\frac{1}{2\sqrt{2}}|^2 + |\frac{1}{2\sqrt{2}}|^2 = \frac{1}{4}$, then we have

$$\frac{\frac{1}{2\sqrt{2}}|\Phi^+\rangle_{A'B'}|00\rangle_{AB} + \frac{1}{2\sqrt{2}}|\Phi^+\rangle_{A'B'}|11\rangle_{AB}}{\sqrt{\frac{1}{4}}} = |\Phi^+\rangle_{A'B'} \otimes \frac{1}{\sqrt{2}}(|00\rangle + |11\rangle)_{AB}$$

(see Chapter 2 on measurement) and the qubits A and B share the entangled state $\frac{1}{\sqrt{2}}(|00\rangle + |11\rangle)$, i.e. the entangled state is now between qubits at nodes 1 and 3! Node 2 between nodes 1 and 3 can be thought of as a quantum repeater "extending the range of the entanglement."

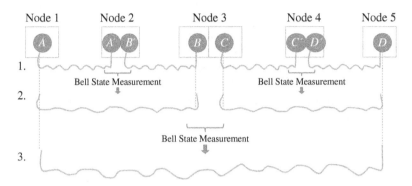

Figure 6.2 Illustration of repeated application of entanglement swapping to generate an entangled pair between qubits A and D at nodes 1 and 5, respectively.

A useful view of entanglement swapping is the teleportation of qubit A' at node 2 (which has been entangled with A at node 1) to node 3 using the entangled pair between B' and B as a resource, i.e. B takes on the state of A', i.e. B is now entangled with A.

A repeated application of such entanglement swapping can be used to create entanglement between a pair of qubits over long distances, as illustrated in Figure 6.2. Although the example shows only five nodes, the same idea can be extended over many more nodes, i.e. across the Internet.

6.1.3 Transmission of Qubits Using Tree-Cluster States

We have looked at entanglement swapping as a way to generate entanglement among nodes far apart. We consider another way here.

We mentioned earlier that one way to transfer quantum information is using teleportation (provided we already have entangled pair(s)) and the other way is by direct transmission. For example, to create an entangled pair between qubits on two nodes geographically far apart, we can, in principle, first locally generate a Bell pair at one node, and then transmit one of the qubits to the other node, and so, have an entangled pair between the two nodes. However, transmitting qubits (e.g. via transmitting photons) over long distances can result in losses (e.g. due to attenuation in a fiber-optic channel). One way to deal with this is to use repeater stations between source and destinations nodes. But such repeaters, due to the no-cloning theorem, cannot make copies of arbitrary unknown quantum states (like in the classical case), but we outline here a method to encode information robustly to allow robust transmission and also allow the qubit information to be re-encoded at repeaters to be transmitted further on, due to Borregaard et al. [2020]. This method also allows transmission of qubit information without pre-established entangled links (unlike the quantum teleportation protocol which, we recall from Chapter 2, requires one EPR pair and transmitting two classical bits).

The method is as follows. Recall that the information in a qubit $\alpha |0\rangle + \beta |1\rangle$ are in the amplitudes α and β. Instead of encoding this information in one photon which is transmitted, one can encode this information in a "cluster of photons" (in a particular *tree-cluster state*), which is more robust against transmission losses. For example, consider the 7-qubit tree-cluster state as follows:

$$|\tau\rangle = \frac{1}{2\sqrt{2}} |0\rangle_s \otimes (|0\rangle |++\rangle + |1\rangle |--\rangle) \otimes (|0\rangle |++\rangle + |1\rangle |--\rangle)$$

$$+ \frac{1}{2\sqrt{2}} |1\rangle_s \otimes (|0\rangle |++\rangle - |1\rangle |--\rangle) \otimes (|0\rangle |++\rangle - |1\rangle |--\rangle)$$

where the subscript s denotes a stationary qubit, and the rest will be qubits to be transmitted.[4]

The method is to then do a Bell measurement with the stationary qubit s and the qubit to be transmitted (i.e. the message qubit m) in the following state $(\alpha |0\rangle + \beta |1\rangle)_m \otimes |\tau\rangle$, which effectively "transfers the quantum information from the message qubit to the tree-cluster state," that is, we have:

$$(H \otimes I^{\otimes 7}) CNOT_{ms} (\alpha |0\rangle + \beta |1\rangle)_m \otimes |\tau\rangle$$

$$= (H \otimes I^{\otimes 7}) CNOT_{ms} (\alpha |0\rangle + \beta |1\rangle)_m$$

$$\otimes \left(\frac{1}{2\sqrt{2}} |0\rangle_s \otimes (|0\rangle |++\rangle + |1\rangle |--\rangle) \otimes (|0\rangle |++\rangle + |1\rangle |--\rangle) \right.$$

$$\left. + \frac{1}{2\sqrt{2}} |1\rangle_s \otimes (|0\rangle |++\rangle - |1\rangle |--\rangle) \otimes (|0\rangle |++\rangle - |1\rangle |--\rangle) \right)$$

---

4 It is called a *tree*-cluster state as one can view the 7-qubits as a tree with the first qubit being the root (s), and then the second and fifth qubits (called first-level qubits) being the child of the root ($s.1$ and $s.2$), and the third and fourth qubits (called second-level qubits) being children of the second qubit ($s.1.1, s.1.2$), and the last two qubits (also called second-level qubits) as children of the fifth qubit ($s.2.1, s.2.2$):

$$|\tau\rangle = \frac{1}{2\sqrt{2}} |0\rangle_s \otimes (|0\rangle_{s.1} |++\rangle_{s.1.1,s.1.2} + |1\rangle_{s.1} |--\rangle_{s.1.1,s.1.2})$$

$$\otimes (|0\rangle_{s.2} |++\rangle_{s.2.1,s.2.2} + |1\rangle_{s.2} |--\rangle_{s.2.1,s.2.2})$$

$$+ \frac{1}{2\sqrt{2}} |1\rangle_s \otimes (|0\rangle_{s.1} |++\rangle_{s.1.1,s.1.2} - |1\rangle_{s.1} |--\rangle_{s.1.1,s.1.2})$$

$$\otimes (|0\rangle_{s.2} |++\rangle_{s.2.1,s.2.2} - |1\rangle_{s.2} |--\rangle_{s.2.1,s.2.2})$$

And letting $f_1 = (|0\rangle\,|++\rangle + |1\rangle\,|--\rangle)^{\otimes 2}$ and $f_2 = (|0\rangle\,|++\rangle - |1\rangle\,|--\rangle)^{\otimes 2}$, we have the above equal to:

$$(H \otimes I^{\otimes 7})CNOT_{ms}\,(\alpha\,|0\rangle + \beta\,|1\rangle)_m \otimes \left(\frac{1}{2\sqrt{2}}|0\rangle_s \otimes f_1 + \frac{1}{2\sqrt{2}}|1\rangle_s \otimes f_2 \right)$$

$$= (H \otimes I^{\otimes 7})CNOT_{ms} \left[\frac{\alpha}{2\sqrt{2}}|00\rangle_{ms} \otimes f_1 + \frac{\alpha}{2\sqrt{2}}|01\rangle_{ms} \otimes f_2 \right.$$

$$\left. + \frac{\beta}{2\sqrt{2}}|10\rangle_{ms} \otimes f_1 + \frac{\beta}{2\sqrt{2}}|11\rangle_{ms} \otimes f_2 \right]$$

$$= (H \otimes I^{\otimes 7}) \left[\frac{\alpha}{2\sqrt{2}}|00\rangle_{ms} \otimes f_1 + \frac{\alpha}{2\sqrt{2}}|01\rangle_{ms} \otimes f_2 \right.$$

$$\left. + \frac{\beta}{2\sqrt{2}}|11\rangle_{ms} \otimes f_1 + \frac{\beta}{2\sqrt{2}}|10\rangle_{ms} \otimes f_2 \right]$$

$$= \frac{\alpha}{4}(|00\rangle + |10\rangle)_{ms} \otimes f_1 + \frac{\alpha}{4}(|01\rangle + |11\rangle)_{ms} \otimes f_2$$

$$+ \frac{\beta}{4}(|01\rangle - |11\rangle)_{ms} \otimes f_1 + \frac{\beta}{4}(|00\rangle - |10\rangle)_{ms} \otimes f_2$$

$$= |00\rangle_{ms} \otimes \left(\frac{\alpha}{4}f_1 + \frac{\beta}{4}f_2 \right) + |01\rangle_{ms} \otimes \left(\frac{\beta}{4}f_1 + \frac{\alpha}{4}f_2 \right)$$

$$+ |10\rangle_{ms} \otimes \left(\frac{\alpha}{4}f_1 - \frac{\beta}{4}f_2 \right) + |11\rangle_{ms} \otimes \left(-\frac{\beta}{4}f_1 + \frac{\alpha}{4}f_2 \right)$$

where we get $\frac{\alpha}{4}f_1 + \frac{\beta}{4}f_2$, $\frac{\beta}{4}f_1 + \frac{\alpha}{4}f_2$, $\frac{\alpha}{4}f_1 - \frac{\beta}{4}f_2$ or $-\frac{\beta}{4}f_1 + \frac{\alpha}{4}f_2$ (with equal probability, i.e. 1/4) depending on the outcome of the measurement, i.e. when we measure the qubits m and s with respect to the computational basis. Suppose the measurement yielded $|00\rangle_{ms}$, then we transmit the 6-qubit cluster state:

$$|\tau'\rangle = \frac{\frac{\alpha}{4}f_1 + \frac{\beta}{4}f_2}{\sqrt{1/4}} = \frac{\alpha}{2}f_1 + \frac{\beta}{2}f_2, \text{ which is}$$

$$\frac{\alpha}{2}(|0\rangle\,|++\rangle + |1\rangle\,|--\rangle) \otimes (|0\rangle\,|++\rangle + |1\rangle\,|--\rangle)$$

$$+ \frac{\beta}{2}(|0\rangle\,|++\rangle - |1\rangle\,|--\rangle) \otimes (|0\rangle\,|++\rangle - |1\rangle\,|--\rangle)$$

containing the quantum information to be transmitted (i.e. the amplitudes α and β). How is this more robust to transmission losses? From the example in Borregaard et al. [2020], suppose some of the qubits are lost during the transmission, say the first two qubits, we will then be left with either:

$$\text{(i)} \quad \frac{\alpha}{\sqrt{2}}\, |+\rangle \otimes (|0\rangle\, |{++}\rangle + |1\rangle\, |{--}\rangle) + \frac{\beta}{\sqrt{2}}\, |+\rangle \otimes (|0\rangle\, |{++}\rangle - |1\rangle\, |{--}\rangle)$$

or

$$\text{(ii)} \quad \frac{\alpha}{\sqrt{2}}\, |-\rangle \otimes (|0\rangle\, |{++}\rangle + |1\rangle\, |{--}\rangle) - \frac{\beta}{\sqrt{2}}\, |-\rangle \otimes (|0\rangle\, |{++}\rangle - |1\rangle\, |{--}\rangle)$$

with equal probability, since it is like measuring out the first two qubits.[5] Then, we can measure (in the X basis) the first qubit above to get $|+\rangle$ or $|-\rangle$, i.e. one of the two cases. Suppose we get the resulting state in case (i), which we denote by $|(i)\rangle$:

$$|(i)\rangle = \frac{\alpha}{\sqrt{2}}(|0\rangle\, |{++}\rangle + |1\rangle\, |{--}\rangle) + \frac{\beta}{\sqrt{2}}(|0\rangle\, |{++}\rangle - |1\rangle\, |{--}\rangle)$$

$$= \frac{\alpha + \beta}{\sqrt{2}}\, |0\rangle\, |{++}\rangle + \frac{\alpha - \beta}{\sqrt{2}}\, |1\rangle\, |{--}\rangle$$

$$= \frac{\alpha + \beta}{2\sqrt{2}}\, |0\rangle\, (|00\rangle + |01\rangle + |10\rangle + |11\rangle) + \frac{\alpha - \beta}{2\sqrt{2}}\, |1\rangle\, (|00\rangle - |01\rangle - |10\rangle + |11\rangle)$$

(recalling that $|+\rangle = (|0\rangle + |1\rangle)/\sqrt{2}$ and $|-\rangle = (|0\rangle - |1\rangle)/\sqrt{2}$)

$$= \left(\frac{\alpha + \beta}{2\sqrt{2}}\, |0\rangle + \frac{\alpha - \beta}{2\sqrt{2}}\, |1\rangle \right) |00\rangle + \left(\frac{\alpha + \beta}{2\sqrt{2}}\, |0\rangle - \frac{\alpha - \beta}{2\sqrt{2}}\, |1\rangle \right) |01\rangle$$

$$+ \left(\frac{\alpha + \beta}{2\sqrt{2}}\, |0\rangle - \frac{\alpha - \beta}{2\sqrt{2}}\, |1\rangle \right) |10\rangle + \left(\frac{\alpha + \beta}{2\sqrt{2}}\, |0\rangle + \frac{\alpha - \beta}{2\sqrt{2}}\, |1\rangle \right) |11\rangle$$

Then, upon measuring the last two qubits in the Z basis, we get with probability $1/4$ one of the following: $|00\rangle$, $|01\rangle$, $|10\rangle$, or $|11\rangle$. Suppose we get $|00\rangle$, then we obtain:

$$\frac{(I \otimes |0\rangle \langle 0| \otimes |0\rangle \langle 0|)\, |(i)\rangle}{\sqrt{1/4}} = \left(\frac{\alpha + \beta}{\sqrt{2}}\, |0\rangle + \frac{\alpha - \beta}{\sqrt{2}}\, |1\rangle \right) \otimes |00\rangle$$

Noting that $\frac{\alpha + \beta}{\sqrt{2}}\, |0\rangle + \frac{\alpha - \beta}{\sqrt{2}}\, |1\rangle = \alpha\, |+\rangle + \beta\, |-\rangle$, by applying the Hadamard (H) gate to $\alpha\, |+\rangle + \beta\, |-\rangle$, we obtain the message qubit $\alpha\, |0\rangle + \beta\, |1\rangle$. If we get $|01\rangle$, we obtain $\frac{\alpha + \beta}{\sqrt{2}}\, |0\rangle - \frac{\alpha - \beta}{\sqrt{2}}\, |1\rangle$, which can be rewritten:

$$\frac{\alpha + \beta}{\sqrt{2}}\, |0\rangle - \frac{\alpha - \beta}{\sqrt{2}}\, |1\rangle = Z(\alpha\, |+\rangle + \beta\, |-\rangle)$$

By applying HZ to $Z(\alpha\, |+\rangle + \beta\, |-\rangle)$, we can then obtain $\alpha\, |0\rangle + \beta\, |1\rangle$, and similarly, with $|10\rangle$ and $|11\rangle$. A similar reasoning with case (ii).

---

5 To get (i), for example, we compute $\frac{(|0\rangle\langle 0|\otimes|+\rangle\langle +|\otimes I^{\otimes 4})\,|\tau'\rangle}{\sqrt{1/2}}$ and leave out writing the measured qubits $|0\rangle\, |+\rangle$. Similarly, if measuring the first two qubits yields $|1\rangle\, |-\rangle$.

Hence, we see that losing two (or fewer) out of the six qubits (in the tree-cluster state) is fine as the quantum state of the message can still be recovered. (A larger tree-cluster state can be used to provide greater robustness.)

The above scheme is a process for recovering the message qubit. The method is basically as follows:

1. Measure out all the qubits in one branch (either one): measure the first-level qubit in the Z basis and second-level qubits in the X basis. (Note that in the example above, the two lost qubits are as though they have been "measured out" and then we measure one more second-level qubit (in that branch where the two qubits were lost) in the X basis, i.e. we measured out all the qubits in one branch.)
2. In the remaining branch, measure the two second-level qubits in the Z basis and perform some H and/or Pauli operations as required.

Once recovered, at a repeater, the qubit can be re-encoded in a similar way with a (new) tree-cluster state and transmitted to the next repeater (or destination).

However, we might want to first "test" to see if there is photon (qubit) loss or not, i.e. if the encoded message qubit arrived successfully or not, before proceeding to re-encoding with a new tree-cluster state. This is done as follows. Suppose the qubits received at this repeater is $|\tau'\rangle = \frac{\alpha}{2}f_1 + \frac{\beta}{2}f_2$, i.e.

$$|\tau'\rangle = \frac{\alpha}{2}(|0\rangle_{c1}|++\rangle + |1\rangle_{c1}|--\rangle) \otimes (|0\rangle_{c2}|++\rangle + |1\rangle_{c2}|--\rangle)$$

$$+ \frac{\beta}{2}(|0\rangle_{c1}|++\rangle - |1\rangle_{c1}|--\rangle) \otimes (|0\rangle_{c2}|++\rangle - |1\rangle_{c2}|--\rangle)$$

where we highlight the first-level qubits (recalling the tree-cluster idea where the first level qubits are marked as $c1$ and $c2$ here). The controlled-Z gate (CZ) operation is done with one of the first-level qubits in the transmitted state (say $c1$) and an auxiliary stationary qubit s in the state $H|0\rangle$ and a measurement in the X basis is done. If the X basis measurement was successful, then a transfer to the new cluster state is *heralded*, and a Bell measurement is done between the auxiliary qubit and the root of a new tree-cluster state, transferring the quantum information to the new tree-cluster state. The circuit (showing only qubits $c1$ and s) is as follows:

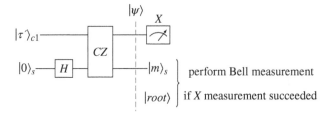

In the above circuit, $|m\rangle_s$ denotes the state of the auxiliary qubit just before the Bell measurement, and $|\psi\rangle$ is used here to denote the state of the c1 and s qubits after the CZ gate, both of which we detail below.

The received encoded qubit (i.e. a cluster of qubits) τ', and the auxiliary qubit s can be written as follows:

$$|\tau'\rangle \otimes |0\rangle_s = \frac{\alpha}{2}|0\rangle_{c1}|++\rangle \otimes (|0\rangle_{c2}|++\rangle + |1\rangle_{c2}|--\rangle) \otimes |0\rangle_s$$

$$+ \frac{\alpha}{2}|1\rangle_{c1}|--\rangle \otimes (|0\rangle_{c2}|++\rangle + |1\rangle_{c2}|--\rangle) \otimes |0\rangle_s$$

$$+ \frac{\beta}{2}|0\rangle_{c1}|++\rangle \otimes (|0\rangle_{c2}|++\rangle - |1\rangle_{c2}|--\rangle) \otimes |0\rangle_s$$

$$- \frac{\beta}{2}|1\rangle_{c1}|--\rangle \otimes (|0\rangle_{c2}|++\rangle - |1\rangle_{c2}|--\rangle) \otimes |0\rangle_s$$

Then, looking at the circuit, after applying the H gate to $|0\rangle_s$ and $CZ_{c1,s}$, we get the state $|\psi\rangle$:

$$|\psi\rangle = \frac{\alpha}{2}|0\rangle_{c1}|++\rangle \otimes (|0\rangle_{c2}|++\rangle + |1\rangle_{c2}|--\rangle) \otimes \frac{1}{\sqrt{2}}(|0\rangle + |1\rangle)_s$$

$$+ \frac{\alpha}{2}|1\rangle_{c1}|--\rangle \otimes (|0\rangle_{c2}|++\rangle + |1\rangle_{c2}|--\rangle) \otimes \frac{1}{\sqrt{2}}(|0\rangle - |1\rangle)_s$$

$$+ \frac{\beta}{2}|0\rangle_{c1}|++\rangle \otimes (|0\rangle_{c2}|++\rangle - |1\rangle_{c2}|--\rangle) \otimes \frac{1}{\sqrt{2}}(|0\rangle + |1\rangle)_s$$

$$- \frac{\beta}{2}|1\rangle_{c1}|--\rangle \otimes (|0\rangle_{c2}|++\rangle - |1\rangle_{c2}|--\rangle) \otimes \frac{1}{\sqrt{2}}(|0\rangle - |1\rangle)_s$$

Then, as in the circuit, an X basis measurement is done; suppose the X measurement is successful, and $|+\rangle$ is obtained, the above state becomes: $|\psi'\rangle = \dfrac{|+\rangle\langle+| \otimes I^{\otimes 6}|\psi\rangle}{\sqrt{1/2}}$, that is (recalling that $|0\rangle = (|+\rangle + |-\rangle)/\sqrt{2}$ and $|1\rangle = (|+\rangle - |-\rangle)/\sqrt{2}$):

$$|\psi'\rangle = \frac{\alpha}{2}|+\rangle_{c1}|++\rangle \otimes (|0\rangle_{c2}|++\rangle + |1\rangle_{c2}|--\rangle) \otimes \frac{1}{\sqrt{2}}(|0\rangle + |1\rangle)_s$$

$$+ \frac{\alpha}{2}|+\rangle_{c1}|--\rangle \otimes (|0\rangle_{c2}|++\rangle + |1\rangle_{c2}|--\rangle) \otimes \frac{1}{\sqrt{2}}(|0\rangle - |1\rangle)_s$$

$$+ \frac{\beta}{2}|+\rangle_{c1}|++\rangle \otimes (|0\rangle_{c2}|++\rangle - |1\rangle_{c2}|--\rangle) \otimes \frac{1}{\sqrt{2}}(|0\rangle + |1\rangle)_s$$

$$- \frac{\beta}{2}|+\rangle_{c1}|--\rangle \otimes (|0\rangle_{c2}|++\rangle - |1\rangle_{c2}|--\rangle) \otimes \frac{1}{\sqrt{2}}(|0\rangle - |1\rangle)_s$$

which can be simplified, including leaving out the measured first qubit $c1$, so that we now write:

$$|\psi'\rangle = \frac{\alpha}{2}|++\rangle \otimes (|0\rangle_{c2}|++\rangle + |1\rangle_{c2}|--\rangle) \otimes |+\rangle_s$$

$$+ \frac{\alpha}{2}|--\rangle \otimes (|0\rangle_{c2}|++\rangle + |1\rangle_{c2}|--\rangle) \otimes |-\rangle_s$$

$$+ \frac{\beta}{2}|++\rangle \otimes (|0\rangle_{c2}|++\rangle - |1\rangle_{c2}|--\rangle) \otimes |+\rangle_s$$

$$- \frac{\beta}{2}|--\rangle \otimes (|0\rangle_{c2}|++\rangle - |1\rangle_{c2}|--\rangle) \otimes |-\rangle_s$$

Suppose this is successful, then the right branch qubits (i.e. "$(|0\rangle_{c2}|++\rangle + |1\rangle_{c2}|--\rangle)$" above) can be measured out. Let's see what we are then left with. We write the state for $|\psi''\rangle$, leaving out the (measured out) right branch qubits[6]:

$$|\psi''\rangle = \frac{\alpha}{\sqrt{2}}|++\rangle \otimes |+\rangle_s + \frac{\alpha}{\sqrt{2}}|--\rangle \otimes |-\rangle_s$$

$$+ \frac{\beta}{\sqrt{2}}|++\rangle \otimes |+\rangle_s - \frac{\beta}{\sqrt{2}}|--\rangle \otimes |-\rangle_s$$

$$= \frac{\alpha+\beta}{\sqrt{2}}|++\rangle \otimes |+\rangle_s + \frac{\alpha-\beta}{\sqrt{2}}|--\rangle \otimes |-\rangle_s$$

$$= \frac{\alpha+\beta}{2\sqrt{2}}(|00\rangle + |01\rangle + |10\rangle + |11\rangle) \otimes |+\rangle_s$$

$$+ \frac{\alpha-\beta}{2\sqrt{2}}(|00\rangle - |01\rangle - |10\rangle + |11\rangle) \otimes |-\rangle_s$$

$$= |00\rangle \otimes \left(\frac{\alpha+\beta}{2\sqrt{2}}|+\rangle_s + \frac{\alpha-\beta}{2\sqrt{2}}|-\rangle_s\right) + |01\rangle \otimes \left(\frac{\alpha+\beta}{2\sqrt{2}}|+\rangle_s - \frac{\alpha-\beta}{2\sqrt{2}}|-\rangle_s\right)$$

$$+ |10\rangle \otimes \left(\frac{\alpha+\beta}{2\sqrt{2}}|+\rangle_s - \frac{\alpha-\beta}{2\sqrt{2}}|-\rangle_s\right) + |11\rangle \otimes \left(\frac{\alpha+\beta}{2\sqrt{2}}|+\rangle_s + \frac{\alpha-\beta}{2\sqrt{2}}|-\rangle_s\right)$$

Now we measure the first two qubits shown above, and suppose the measurement of these second-level qubits (associated with $c1$) in the Z basis yielded $|00\rangle$ with probability $1/4$, we obtain $\frac{1}{\sqrt{1/4}}(\frac{\alpha+\beta}{2\sqrt{2}}|+\rangle_s + \frac{\alpha-\beta}{2\sqrt{2}}|-\rangle_s) = \frac{\alpha+\beta}{\sqrt{2}}|+\rangle_s + \frac{\alpha-\beta}{\sqrt{2}}|-\rangle_s = \alpha|0\rangle_s + \beta|1\rangle_s$, i.e. $|m\rangle_s = \alpha|0\rangle_s + \beta|1\rangle_s$, which is used in the subsequent Bell measurement

---

6 Note that measuring out the right branch qubits, suppose the measurement yielded $|0\rangle_{c2}|++\rangle$ (with probability $1/2$), we then have the state: $\dfrac{(I \otimes I \otimes |0++\rangle\langle 0++| \otimes I)|\psi'\rangle}{\sqrt{1/2}}$.

to be encoded in the new tree-cluster state. This is similar if $|11\rangle$ was obtained. But if the measurement yielded $|01\rangle$, we obtain $\beta|0\rangle_s + \alpha|1\rangle_s$, and then applying an X gate would yield $\alpha|0\rangle_s + \beta|1\rangle_s$ which can be used in the Bell measurement (and similarly, for $|10\rangle$). Note that this is similar to how we dealt with $|(i)\rangle$ earlier.

Note that if measuring $c1$ is not successful (i.e. the X measurement failed), then the same process is retried for the other first-level qubit, $c2$. A Bell measurement is then done with the qubit denoted by $|m\rangle_s$ in the circuit and the root of a new tree-cluster state, i.e. reencoding the state in a new tree-cluster state for further transmission.

Three generations of quantum repeaters have been identified in Yan et al. [2021] and Muralidharan et al. [2016], and the above illustrates the idea of a third-generation quantum repeater, where "redundant" qubits are used when transmitting entangled qubits (e.g. one qubit from a pair of entangled qubits) in order to be fault tolerant.

6.2 Entanglement with Higher Fidelity

Decoherence due to noise was mentioned earlier which limits the distances over which entanglement might be generated; entanglement swapping provided a way to entangle qubits much further apart over long distances. However, there is still a need to improve the "quality" of entanglement, that is, to improve the fidelity of qubit states toward the desired ideal. Before introducing a method to improve fidelity, we first define fidelity more precisely and also introduce the notion of the twirling map.

6.2.1 Fidelity

We first consider a precise model of fidelity, of what it means for a quantum state to be close to some ideal.[7] Due to noise, an entangled state may not be what exactly we want it to be. The consequences of noisy distributed entanglement include failure of particular nonlocal quantum gate operations and problems in quantum protocols, e.g. in QKD, an imperfect entangled state produces a partially correlated key. There is a need to evaluate the quality of a distributed entanglement and then think of how to improve the quality of the entanglement. One way to measure the quality of entanglement is to use the notion of *fidelity*. Given a reference (ideal) state $|\psi\rangle$ and the actual state ρ shared by two nodes, the fidelity, denoted by the function F, between the reference and actual state is $F(\rho, |\psi\rangle) = \langle\psi|\rho|\psi\rangle$. There are

---

7 Our discussions are based on the lectures at https://www.youtube.com/watch?v=MD9uDrly4Ms and https://www.youtube.com/watch?v=G5z6a8Mp-9w [last accessed: 16/11/2022].

other definitions of fidelity but we use the one as just given for the purposes of our discussion here.

Note suppose $|\psi\rangle = \alpha\,|0\rangle + \beta\,|1\rangle$, then its density operator representation

$$\rho = |\psi\rangle\langle\psi| = (\alpha\,|0\rangle + \beta\,|1\rangle)(\alpha^*\,\langle 0| + \beta^*\,\langle 1|)$$
$$= |\alpha|^2\,|0\rangle\langle 0| + \alpha\beta^*\,|0\rangle\langle 1| + \beta\alpha^*\,|1\rangle\langle 0| + |\beta|^2\,|1\rangle\langle 1|$$

where $(\cdot)^*$ is the complex conjugate. That is, the probability of $|\psi\rangle$ being found to be $|0\rangle$ is $\langle 0|\rho|0\rangle = F(\rho, |0\rangle) = |\alpha|^2$. Also, the probability of $|\psi\rangle$ being found to be $|1\rangle$ is $|\beta|^2$. With a two qubit maximally mixed state represented by the density operator:

$$\rho = \frac{1}{4}(|00\rangle\langle 00| + |01\rangle\langle 01| + |10\rangle\langle 10| + |11\rangle\langle 11|)$$

and $|\psi\rangle = |00\rangle$, we have: $F(\rho, |\psi\rangle) = \langle 00|\rho|00\rangle = \frac{1}{4}$. For an N qubit state, we have: $F(\rho, |00\rangle) = \frac{1}{2^N}$.

Fidelity measures the quality of the actual state with respect to the reference state and can be interpreted as the probability of the actual state being detected as the reference state (or the "overlap" between the actual and reference states).

With respect to $|\Phi^+\rangle$, we have: $F(\rho, |\Phi^+\rangle) = \langle \Phi^+|\rho|\Phi^+\rangle$. With no noise, suppose $\rho = |\Phi^+\rangle\langle \Phi^+|$, then: $F(\rho, |\Phi^+\rangle) = \langle \Phi^+|\Phi^+\rangle\langle \Phi^+|\Phi^+\rangle = 1$.

Consider the flip channel modeling some noise, with probability of a bit flip being p and at most only one bit flip can occur, where for a one qubit input state $|\psi\rangle$, the output is $\rho = (1-p)\,|\psi\rangle\langle\psi| + pX\,|\psi\rangle\langle\psi|\,X$, Then,

$$F(\rho, |\psi\rangle) = \langle\psi|\,((1-p)\,|\psi\rangle\langle\psi| + pX\,|\psi\rangle\langle\psi|\,X)\,|\psi\rangle$$
$$= (1-p)\,\langle\psi|\psi\rangle\langle\psi|\psi\rangle + p\,\langle\psi|X|\psi\rangle\langle\psi|X|\psi\rangle$$
$$= 1 - p(1 - \langle\psi|X|\psi\rangle^2) \leq 1$$

which has the potential to decrease the fidelity.

As another example, suppose we desire state $|\Phi^+\rangle$ (or as the density matrix $|\Phi^+\rangle\langle \Phi^+|$), but the actual state ρ might be "impure" in that it has some probability p of being in an unknown noisy state $|unknown\rangle\langle unknown|$:

$$\rho = (1-p)\,|\Phi^+\rangle\langle \Phi^+| + p\,|unknown\rangle\langle unknown|$$

Then,

$$F(\rho, |\Phi^+\rangle) = \langle \Phi^+|\,((1-p)\,|\Phi^+\rangle\langle \Phi^+| + p\,|unknown\rangle\langle unknown|)\,|\Phi^+\rangle$$
$$= 1 - p + p(\langle unknown|\Phi^+\rangle)^2 \leq 1$$

Note that one can use the CHSH game seen in Chapter 3 to test the "purity" of entangled pairs. Recall that we made use of the state $|\Phi^-\rangle$ and applied the operations as follows: $(R(\theta_1) \otimes R(\theta_2)) \frac{1}{\sqrt{2}} (|00\rangle - |11\rangle) = \frac{1}{\sqrt{2}} (\cos(\theta_1 + \theta_2)(|00\rangle - |11\rangle) + \sin(\theta_1 + \theta_2)(|01\rangle + |10\rangle))$, or equivalently:

$$(R(\theta_1) \otimes R(\theta_2)) |\Phi^-\rangle = \cos(\theta_1 + \theta_2) |\Phi^-\rangle + \sin(\theta_1 + \theta_2) |\Psi^+\rangle$$

In the presence of noise modeled earlier, we would then have:

$$(R(\theta_1) \otimes R(\theta_2)) \left(\sqrt{1-p} |\Phi^-\rangle + \sqrt{p} |unknown'\rangle \right)$$
$$= \sqrt{1-p}[\cos(\theta_1 + \theta_2) |\Phi^-\rangle + \sin(\theta_1 + \theta_2) |\Psi^+\rangle] + \sqrt{p} |noise\rangle$$

where $|noise\rangle$ represents the operations on the unknown state. This means that, with suitable θ_1 and θ_2 as in Chapter 3, measurement would yield $|\Phi^-\rangle$ with probability $(1-p)\cos^2(\pi/8) < 0.8536$ with large enough p, and with large enough p, this probability might go to around 0.75 or below, which is no longer better than the best classical case (of 0.75)! This means that (the rate or probability of) winning the CHSH game (i.e. more than probability 0.75) can be a "test" that we do have (or are achieving) a high fidelity entanglement.

6.2.2 Twirling Map

We will also need to use the following concept called a *twirling map*. We define a twirling map $\tau : D(\mathbb{C}^2 \otimes \mathbb{C}^2) \to D(\mathbb{C}^2 \otimes \mathbb{C}^2)$, where $D(\mathbb{C}^2 \otimes \mathbb{C}^2)$ refers to the set of two qubit density operators, that maps any bipartite density operator ρ_{AB} to a *Bell diagonal state*, of the following form:

$$p_1 |\Phi^+\rangle \langle\Phi^+| + p_2 |\Phi^-\rangle \langle\Phi^-| + p_3 |\Psi^+\rangle \langle\Psi^+| + p_4 |\Psi^-\rangle \langle\Psi^-|$$

where $\sum_{i=1}^{4} p_i = 1$. It has the property of being a completely positive trace-preserving (CPTP) map.[8]

The twirling map picks a Pauli operator uniformly at random (with equal probability) and applies it to each qubit in ρ_{AB}; we write it in the form:

$$\tau(\rho_{AB}) = \frac{1}{4} \sum_i (E_i \otimes E_i) \rho_{AB} (E_i \otimes E_i)^\dagger$$

where $E_i \in \{I, X, Y, Z\}$, i.e. the E_i are the Pauli operators (e.g. $E_0 = I$, $E_1 = X$, $E_2 = Y$, and $E_3 = Z$).

To see how the twirling map works, we write a state $|\psi\rangle$ in terms of the Bell basis:

$$|\psi\rangle = \alpha_1 |\Phi^+\rangle + \alpha_2 |\Phi^-\rangle + \alpha_3 |\Psi^+\rangle + \alpha_4 |\Psi^-\rangle$$

---

8 For more on a CPTP map, see Artur Ekert and Tim Hosgood's book online: https://qubit .guide/9.9-completely-positive-trace-preserving-maps [last accessed: 1/10/2023].

where $\sum_i |\alpha_i|^2 = 1$. Let $\rho_{AB} = |\psi\rangle \langle \psi|$, i.e.

$$\rho_{AB} = \left(\alpha_1 |\Phi^+\rangle + \alpha_2 |\Phi^-\rangle + \alpha_3 |\Psi^+\rangle + \alpha_4 |\Psi^-\rangle \right)$$
$$\times (\alpha_1^* \langle \Phi^+| + \alpha_2^* \langle \Phi^-| + \alpha_3^* \langle \Psi^+| + \alpha_4^* \langle \Psi^-|)$$

which is

$$\rho_{AB}\rho_{AB} = \alpha_1 \alpha_1^* |\Phi^+\rangle \langle \Phi^+| + \alpha_1 \alpha_2^* |\Phi^+\rangle \langle \Phi^-| + \alpha_1 \alpha_3^* |\Phi^+\rangle \langle \Psi^+| + \alpha_1 \alpha_4^* |\Phi^+\rangle \langle \Psi^-|$$
$$+ \alpha_2 \alpha_1^* |\Phi^-\rangle \langle \Phi^+| + \alpha_2 \alpha_2^* |\Phi^-\rangle \langle \Phi^-| + \alpha_2 \alpha_3^* |\Phi^-\rangle \langle \Psi^+| + \alpha_2 \alpha_4^* |\Phi^-\rangle \langle \Psi^-|$$
$$+ \alpha_3 \alpha_1^* |\Psi^+\rangle \langle \Phi^+| + \alpha_3 \alpha_2^* |\Psi^+\rangle \langle \Phi^-| + \alpha_3 \alpha_3^* |\Psi^+\rangle \langle \Psi^+| + \alpha_3 \alpha_4^* |\Psi^+\rangle \langle \Psi^-|$$
$$+ \alpha_4 \alpha_1^* |\Psi^-\rangle \langle \Phi^+| + \alpha_4 \alpha_2^* |\Psi^-\rangle \langle \Phi^-| + \alpha_4 \alpha_3^* |\Psi^-\rangle \langle \Psi^+| + \alpha_4 \alpha_4^* |\Psi^-\rangle \langle \Psi^-|$$

We note that since $X |0\rangle = |1\rangle$, $X |1\rangle = |0\rangle$, $Z |0\rangle = |0\rangle$, $Z |1\rangle = -|1\rangle$, $Y |0\rangle = i |1\rangle$, and $Y |1\rangle = -i |0\rangle$, we have:

$$
\begin{array}{lll}
(X \otimes X) |\Phi^+\rangle = |\Phi^+\rangle & (Y \otimes Y) |\Phi^+\rangle = -|\Phi^+\rangle & (Z \otimes Z) |\Phi^+\rangle = |\Phi^+\rangle \\
(X \otimes X) |\Phi^-\rangle = -|\Phi^-\rangle & (Y \otimes Y) |\Phi^-\rangle = |\Phi^-\rangle & (Z \otimes Z) |\Phi^-\rangle = |\Phi^-\rangle \\
(X \otimes X) |\Psi^+\rangle = |\Psi^+\rangle & (Y \otimes Y) |\Psi^+\rangle = |\Psi^+\rangle & (Z \otimes Z) |\Psi^+\rangle = -|\Psi^+\rangle \\
(X \otimes X) |\Psi^-\rangle = -|\Psi^-\rangle & (Y \otimes Y) |\Psi^-\rangle = -|\Psi^-\rangle & (Z \otimes Z) |\Psi^-\rangle = -|\Psi^-\rangle
\end{array}
$$

Since $((X \otimes X) |\Phi^+\rangle)^\dagger = \langle \Phi^+| (X \otimes X)^\dagger = \langle \Phi^+|$, there is a corresponding set of equalities for the $(E_i \otimes E_i)^\dagger$ operators.

We then have $(I \otimes I)\rho_{AB}(I \otimes I) = \rho_{AB}$ and:

$$(X \otimes X)\rho_{AB}(X \otimes X)$$
$$= \alpha_1 \alpha_1^* |\Phi^+\rangle \langle \Phi^+| - \alpha_1 \alpha_2^* |\Phi^+\rangle \langle \Phi^-| + \alpha_1 \alpha_3^* |\Phi^+\rangle \langle \Psi^+| - \alpha_1 \alpha_4^* |\Phi^+\rangle \langle \Psi^-|$$
$$- \alpha_2 \alpha_1^* |\Phi^-\rangle \langle \Phi^+| + \alpha_2 \alpha_2^* |\Phi^-\rangle \langle \Phi^-| - \alpha_2 \alpha_3^* |\Phi^-\rangle \langle \Psi^+| + \alpha_2 \alpha_4^* |\Phi^-\rangle \langle \Psi^-|$$
$$+ \alpha_3 \alpha_1^* |\Psi^+\rangle \langle \Phi^+| - \alpha_3 \alpha_2^* |\Psi^+\rangle \langle \Phi^-| + \alpha_3 \alpha_3^* |\Psi^+\rangle \langle \Psi^+| - \alpha_3 \alpha_4^* |\Psi^+\rangle \langle \Psi^-|$$
$$- \alpha_4 \alpha_1^* |\Psi^-\rangle \langle \Phi^+| + \alpha_4 \alpha_2^* |\Psi^-\rangle \langle \Phi^-| - \alpha_4 \alpha_3^* |\Psi^-\rangle \langle \Psi^+| + \alpha_4 \alpha_4^* |\Psi^-\rangle \langle \Psi^-|$$

and

$$(Y \otimes Y)\rho_{AB}(Y \otimes Y)$$
$$= \alpha_1 \alpha_1^* |\Phi^+\rangle \langle \Phi^+| - \alpha_1 \alpha_2^* |\Phi^+\rangle \langle \Phi^-| - \alpha_1 \alpha_3^* |\Phi^+\rangle \langle \Psi^+| + \alpha_1 \alpha_4^* |\Phi^+\rangle \langle \Psi^-|$$
$$- \alpha_2 \alpha_1^* |\Phi^-\rangle \langle \Phi^+| + \alpha_2 \alpha_2^* |\Phi^-\rangle \langle \Phi^-| + \alpha_2 \alpha_3^* |\Phi^-\rangle \langle \Psi^+| - \alpha_2 \alpha_4^* |\Phi^-\rangle \langle \Psi^-|$$
$$- \alpha_3 \alpha_1^* |\Psi^+\rangle \langle \Phi^+| + \alpha_3 \alpha_2^* |\Psi^+\rangle \langle \Phi^-| + \alpha_3 \alpha_3^* |\Psi^+\rangle \langle \Psi^+| - \alpha_3 \alpha_4^* |\Psi^+\rangle \langle \Psi^-|$$
$$+ \alpha_4 \alpha_1^* |\Psi^-\rangle \langle \Phi^+| - \alpha_4 \alpha_2^* |\Psi^-\rangle \langle \Phi^-| - \alpha_4 \alpha_3^* |\Psi^-\rangle \langle \Psi^+| + \alpha_4 \alpha_4^* |\Psi^-\rangle \langle \Psi^-|$$

and

$$(Z \otimes Z)\rho_{AB}(Z \otimes Z)$$

$$= \alpha_1\alpha_1^* \left|\Phi^+\right\rangle\left\langle\Phi^+\right| + \alpha_1\alpha_2^* \left|\Phi^+\right\rangle\left\langle\Phi^-\right| - \alpha_1\alpha_3^* \left|\Phi^+\right\rangle\left\langle\Psi^+\right| - \alpha_1\alpha_4^* \left|\Phi^+\right\rangle\left\langle\Psi^-\right|$$

$$+ \alpha_2\alpha_1^* \left|\Phi^-\right\rangle\left\langle\Phi^+\right| + \alpha_2\alpha_2^* \left|\Phi^-\right\rangle\left\langle\Phi^-\right| - \alpha_2\alpha_3^* \left|\Phi^-\right\rangle\left\langle\Psi^+\right| - \alpha_2\alpha_4^* \left|\Phi^-\right\rangle\left\langle\Psi^-\right|$$

$$- \alpha_3\alpha_1^* \left|\Psi^+\right\rangle\left\langle\Phi^+\right| - \alpha_3\alpha_2^* \left|\Psi^+\right\rangle\left\langle\Phi^-\right| + \alpha_3\alpha_3^* \left|\Psi^+\right\rangle\left\langle\Psi^+\right| + \alpha_3\alpha_4^* \left|\Psi^+\right\rangle\left\langle\Psi^-\right|$$

$$- \alpha_4\alpha_1^* \left|\Psi^-\right\rangle\left\langle\Phi^+\right| - \alpha_4\alpha_2^* \left|\Psi^-\right\rangle\left\langle\Phi^-\right| + \alpha_4\alpha_3^* \left|\Psi^-\right\rangle\left\langle\Psi^+\right| + \alpha_4\alpha_4^* \left|\Psi^-\right\rangle\left\langle\Psi^-\right|$$

Hence,

$$\tau(\rho_{AB}) = \frac{1}{4}(4\alpha_1\alpha_1^* \left|\Phi^+\right\rangle\left\langle\Phi^+\right| + 4\alpha_2\alpha_2^* \left|\Phi^-\right\rangle\left\langle\Phi^-\right| + 4\alpha_3\alpha_3^* \left|\Psi^+\right\rangle$$

$$\times \left\langle\Psi^+\right| + 4\alpha_4\alpha_4^* \left|\Psi^-\right\rangle\left\langle\Psi^-\right|)$$

$$= \alpha_1\alpha_1^* \left|\Phi^+\right\rangle\left\langle\Phi^+\right| + \alpha_2\alpha_2^* \left|\Phi^-\right\rangle\left\langle\Phi^-\right| + \alpha_3\alpha_3^* \left|\Psi^+\right\rangle\left\langle\Psi^+\right| + \alpha_4\alpha_4^* \left|\Psi^-\right\rangle\left\langle\Psi^-\right|$$

And we also have $\sum_i \alpha_i\alpha_i^* = \sum_i |\alpha_i|^2 = 1$. Note that the effect of τ is to basically "extract" only the "diagonal" components from ρ_{AB}.

Suppose we define another map $\tau' : D(\mathbb{C}^2 \otimes \mathbb{C}^2) \to D(\mathbb{C}^2 \otimes \mathbb{C}^2)$ that chooses uniformly at random an operator R_i from $\{R_1 = X, R_2 = Y, R_3 = Z\}$ and applies $\sqrt{R_i}$ locally to each qubit; we write τ' acting on some state σ_{AB} as follows:

$$\tau'(\sigma_{AB}) = \frac{1}{3}\sum_i \left(\sqrt{R_i} \otimes \sqrt{R_i}\right) \sigma_{AB} \left(\sqrt{R_i} \otimes \sqrt{R_i}\right)^\dagger$$

Note that the $\sqrt{R_i}$ operators can be viewed as $\pi/2$ rotations in the Blog sphere along the corresponding axes (from [Bennett et al., 1996]) with the following effects (see **Aside**). Applying $\sqrt{R_i}$ locally to each qubit interchanges two Bell states and leaves the other two unchanged: $\sqrt{X} \otimes \sqrt{X}$ causes the mapping $\left|\Phi^+\right\rangle \leftrightarrow \left|\Psi^+\right\rangle$, $\sqrt{Y} \otimes \sqrt{Y}$ causes the mapping $\left|\Phi^-\right\rangle \leftrightarrow \left|\Psi^+\right\rangle$ and $\sqrt{Z} \otimes \sqrt{Z}$ does the mapping $\left|\Phi^+\right\rangle \leftrightarrow \left|\Phi^-\right\rangle$.[9]

Aside: It is also noted that the Pauli operator $U \in \{X, Y, Z\}$ and its root $\sqrt{U}$ can be implemented using $\frac{\pi}{2}$ rotations about the axis u, where u is x, y or z, since $\sqrt{U} = e^{\frac{i\pi}{4}}R_u(\frac{\pi}{2})$, where $R_u(\frac{\pi}{2})$ is a $\frac{\pi}{2}$ rotation around axis u of the Bloch sphere [Soeken et al., 2013], i.e. $R_u(\theta) = \cos(\frac{\theta}{2})I - i\sin(\frac{\theta}{2})U$, and $U = \sqrt{U}\sqrt{U} = e^{\frac{i\pi}{4}}R_u(\frac{\pi}{2})e^{\frac{i\pi}{4}}R_u(\frac{\pi}{2}) = e^{\frac{i\pi}{2}}R_u(\pi)$, since $R_u(\theta_1)R_u(\theta_2) = R_u(\theta_1 + \theta_2)$.

---

9 See also the paper on quantum circuits employing roots of the Pauli matrices [Soeken et al., 2013].

Then, letting $\sigma_{AB} = \tau(\rho_{AB})$, from the definition of τ' above, we have[10]:

$$\tau'(\sigma_{AB}) = \tau'(\tau(\rho_{AB})) = \frac{1}{3}\left((\alpha_4\alpha_4^* + \alpha_4\alpha_4^* + \alpha_4\alpha_4^*)\,|\Psi^-\rangle\,\langle\Psi^-|\right.$$

$$+ (\alpha_1\alpha_1^* + \alpha_2\alpha_2^* + \alpha_3\alpha_3^*)\,|\Psi^+\rangle\,\langle\Psi^+|$$

$$+ (\alpha_3\alpha_3^* + \alpha_1\alpha_1^* + \alpha_2\alpha_2^*)\,|\Phi^+\rangle\,\langle\Phi^+|$$

$$\left.+ (\alpha_2\alpha_2^* + \alpha_3\alpha_3^* + \alpha_1\alpha_1^*)\,|\Phi^-\rangle\,\langle\Phi^-|\right)$$

Since $\alpha_1\alpha_1^* + \alpha_2\alpha_2^* + \alpha_3\alpha_3^* + \alpha_4\alpha_4^* = 1$, if we denote $F(\sigma_{AB}, |\Psi^-\rangle) = \langle\Psi^-|\,\sigma_{AB}\,|\Psi^-\rangle$ $= \alpha_4\alpha_4^*$ by F, then the above can be rewritten as

$$\tau'(\sigma_{AB}) = F\,|\Psi^-\rangle\,\langle\Psi^-| + \frac{(1-F)}{3}\,|\Psi^+\rangle\,\langle\Psi^+| + \frac{(1-F)}{3}\,|\Phi^+\rangle$$

$$\times\,\langle\Phi^+| + \frac{(1-F)}{3}\,|\Phi^-\rangle\,\langle\Phi^-|$$

We can compose τ and τ' to get a map $\tau' \circ \tau$.

6.2.3 Quality of Distributed Entanglement and Entanglement Purification

We now come to the idea of *entanglement purification* (or *entanglement distillation*) as a way to get higher fidelity entangled states. The idea is to take two noisy shared entangled states and yield a shared entangled state which might still be noisy but has a higher fidelity, with respect to one of the Bell states, say, than the original two states.

More precisely, we have nodes A and B, starting with two (prepared) copies of a two qubit state, say $\rho_{A_1 B_1}$ and $\rho_{A_2 B_2}$. Suppose we want to get a state with higher fidelity with respect to the singlet state (i.e. $|\Psi^-\rangle$). Assume $F(\rho_{A_1 B_1}, |\Psi^-\rangle) = \langle\Psi^-|\,\rho_{A_1 B_1}\,|\Psi^-\rangle$ is denoted by F. The entanglement distillation procedure is as follows, using a version from Gharibian[11]:

---

10 Note that this is because (an exercise is to check this):

$$(\sqrt{X} \otimes \sqrt{X})\sigma_{AB}(\sqrt{X} \otimes \sqrt{X})^\dagger = \alpha_1\alpha_1^*\,|\Psi^+\rangle\,\langle\Psi^+| + \alpha_2\alpha_2^*\,|\Phi^-\rangle\,\langle\Phi^-| + \alpha_3\alpha_3^*\,|\Phi^+\rangle$$

$$\times\,\langle\Phi^+| + \alpha_4\alpha_4^*\,|\Psi^-\rangle\,\langle\Psi^-|$$

$$(\sqrt{Y} \otimes \sqrt{Y})\sigma_{AB}(\sqrt{Y} \otimes \sqrt{Y})^\dagger = \alpha_1\alpha_1^*\,|\Phi^+\rangle\,\langle\Phi^+| + \alpha_2\alpha_2^*\,|\Psi^+\rangle\,\langle\Psi^+| + \alpha_3\alpha_3^*\,|\Phi^-\rangle$$

$$\times\,\langle\Phi^-| + \alpha_4\alpha_4^*\,|\Psi^-\rangle\,\langle\Psi^-|$$

$$(\sqrt{Z} \otimes \sqrt{Z})\sigma_{AB}(\sqrt{Z} \otimes \sqrt{Z})^\dagger = \alpha_1\alpha_1^*\,|\Phi^-\rangle\,\langle\Phi^-| + \alpha_2\alpha_2^*\,|\Phi^+\rangle\,\langle\Phi^+| + \alpha_3\alpha_3^*\,|\Psi^+\rangle$$

$$\times\,\langle\Psi^+| + \alpha_4\alpha_4^*\,|\Psi^-\rangle\,\langle\Psi^-|$$

11 This version is based on the exposition from https://www.youtube.com/watch? v=sGhaQ5MbiAY and from Bennett et al. [1996]; see also https://groups.uni-paderborn.de/fg-qi/teaching.html [last accessed: 16/11/2022].

1. Apply the map $\tau' \circ \tau$ (we saw in Section 6.2.2) to $\rho_{A_1 B_1}$ and $\rho_{A_2 B_2}$ to get two Bell diagonal states of the form (denoted in this subsection using the symbol σ with subscripts):

$$\sigma_{A_i B_i} = F|\Psi^-\rangle\langle\Psi^-| + \frac{(1-F)}{3}|\Psi^+\rangle\langle\Psi^+| + \frac{(1-F)}{3}|\Phi^+\rangle$$

$$\times \langle\Phi^+| + \frac{(1-F)}{3}|\Phi^-\rangle\langle\Phi^-|$$

for $i = 1,2$.

2. Note that $(I \otimes Y)|\Psi^-\rangle\langle\Psi^-|(I \otimes Y)^\dagger = |\Phi^+\rangle\langle\Phi^+|$.[12] Apply local Y on one qubit (say, B's qubit) to interchange $|\Phi^+\rangle$ and $|\Psi^-\rangle$ to get:

$$\sigma'_{A_i B_i} = \frac{(1-F)}{3}|\Psi^-\rangle\langle\Psi^-| + \frac{(1-F)}{3}|\Psi^+\rangle\langle\Psi^+| + F|\Phi^+\rangle$$

$$\times \langle\Phi^+| + \frac{(1-F)}{3}|\Phi^-\rangle\langle\Phi^-|$$

for $i = 1,2$. Note that the interchange is to "move the probabilities" from the mostly singlet state to the mostly $|\Phi^+\rangle$ state to prepare for the next step using CNOT gates.

3. Apply CNOT gates locally on both A's side and B's side, and both locally measure the second qubit with outputs a and b, respectively.
   We use the following circuit:

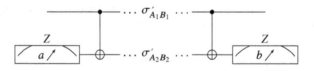

or equivalently:

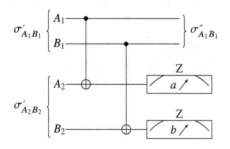

---

12 Note that $Y = i \begin{bmatrix} 0 & -1 \\ 1 & 0 \end{bmatrix}$ so that $Y|0\rangle = i|1\rangle$ and $Y|1\rangle = -i|0\rangle$. Hence, $I \otimes Y|\Phi^+\rangle =$

$I \otimes Y(|00\rangle + |11\rangle)/\sqrt{2} = (i|01\rangle - i|10\rangle)/\sqrt{2} = i|\Psi^-\rangle$, and so $(I \otimes Y|\Phi^+\rangle)(I \otimes Y|\Phi^+\rangle)^\dagger =$
$(i|\Psi^-\rangle)(i|\Psi^-\rangle)^\dagger = (i|\Psi^-\rangle)(-i\langle\Psi^-|) = (-i^2)|\Psi^-\rangle\langle\Psi^-| = |\Psi^-\rangle\langle\Psi^-|$. A similar interchange for the other two Bell states.

Now, if $a = b$, then keep the resulting state $\sigma''_{A_1B_1}$ and keep going to the next step. Otherwise, STOP and start again.

4. Apply local Y on B qubit to interchange back $|\Phi^+\rangle$ and $|\Psi^-\rangle$, i.e. the probabilities are moved back to accumulate on the singlet state.

Let us look at step 3 in more detail. In step 3, Alice and Bob both apply their CNOT gates to two shared pairs (or two copies of σ_{AB}), involving a CNOT involving qubits A_1(as control) and A_2(as target), and a CNOT involving qubits B_1(as control) and B_2(as target), i.e. we apply $CNOT_{A_1A_2} \otimes CNOT_{B_1B_2}$ to $\sigma'_{A_1B_1} \otimes \sigma'_{A_2B_2}$.

To see the effect of step 3, we first compute the following, *ignoring the values of the coefficients c_i for now*:

$$
\begin{aligned}
\sigma'_{A_1B_1} \otimes \sigma'_{A_2B_2} = \ & c_1 |\Psi^-\rangle\langle\Psi^-| \otimes |\Psi^-\rangle\langle\Psi^-| + c_2 |\Psi^-\rangle\langle\Psi^-| \otimes |\Psi^+\rangle\langle\Psi^+| \\
& + c_3 |\Psi^-\rangle\langle\Psi^-| \otimes |\Phi^+\rangle\langle\Phi^+| + c_4 |\Psi^-\rangle\langle\Psi^-| \otimes |\Phi^-\rangle\langle\Phi^-| \\
& + c_5 |\Psi^+\rangle\langle\Psi^+| \otimes |\Psi^-\rangle\langle\Psi^-| + c_6 |\Psi^+\rangle\langle\Psi^+| \otimes |\Psi^+\rangle\langle\Psi^+| \\
& + c_7 |\Psi^+\rangle\langle\Psi^+| \otimes |\Phi^+\rangle\langle\Phi^+| + c_8 |\Psi^+\rangle\langle\Psi^+| \otimes |\Phi^-\rangle\langle\Phi^-| \\
& + c_9 |\Phi^+\rangle\langle\Phi^+| \otimes |\Psi^-\rangle\langle\Psi^-| + c_{10} |\Phi^+\rangle\langle\Phi^+| \otimes |\Psi^+\rangle\langle\Psi^+| \\
& + c_{11} |\Phi^+\rangle\langle\Phi^+| \otimes |\Phi^+\rangle\langle\Phi^+| + c_{12} |\Phi^+\rangle\langle\Phi^+| \otimes |\Phi^-\rangle\langle\Phi^-| \\
& + c_{13} |\Phi^-\rangle\langle\Phi^-| \otimes |\Psi^-\rangle\langle\Psi^-| + c_{14} |\Phi^-\rangle\langle\Phi^-| \otimes |\Psi^+\rangle\langle\Psi^+| \\
& + c_{15} |\Phi^-\rangle\langle\Phi^-| \otimes |\Phi^+\rangle\langle\Phi^+| + c_{16} |\Phi^-\rangle\langle\Phi^-| \otimes |\Phi^-\rangle\langle\Phi^-|
\end{aligned}
$$

and since $AB \otimes CD = (A \otimes C)(B \otimes D) = (AC)(BD)$, we rewrite the above as

$$
\begin{aligned}
\sigma'_{A_1B_1} \otimes \sigma'_{A_2B_2} = \ & c_1 |\Psi^-\rangle|\Psi^-\rangle\langle\Psi^-|\langle\Psi^-| + c_2 |\Psi^-\rangle|\Psi^+\rangle\langle\Psi^-|\langle\Psi^+| \\
& + c_3 |\Psi^-\rangle|\Phi^+\rangle\langle\Psi^-|\langle\Phi^+| + c_4 |\Psi^-\rangle|\Phi^-\rangle\langle\Psi^-|\langle\Phi^-| \\
& + c_5 |\Psi^+\rangle|\Psi^-\rangle\langle\Psi^+|\langle\Psi^-| + c_6 |\Psi^+\rangle|\Psi^+\rangle\langle\Psi^+|\langle\Psi^+| \\
& + c_7 |\Psi^+\rangle|\Phi^+\rangle\langle\Psi^+|\langle\Phi^+| + c_8 |\Psi^+\rangle|\Phi^-\rangle\langle\Psi^+|\langle\Phi^-| \\
& + c_9 |\Phi^+\rangle|\Psi^-\rangle\langle\Phi^+|\langle\Psi^-| + c_{10} |\Phi^+\rangle|\Psi^+\rangle\langle\Phi^+|\langle\Psi^+| \\
& + c_{11} |\Phi^+\rangle|\Phi^+\rangle\langle\Phi^+|\langle\Phi^+| + c_{12} |\Phi^+\rangle|\Phi^-\rangle\langle\Phi^+|\langle\Phi^-| \\
& + c_{13} |\Phi^-\rangle|\Psi^-\rangle\langle\Phi^-|\langle\Psi^-| + c_{14} |\Phi^-\rangle|\Psi^+\rangle\langle\Phi^-|\langle\Psi^+| \\
& + c_{15} |\Phi^-\rangle|\Phi^+\rangle\langle\Phi^-|\langle\Phi^+| + c_{16} |\Phi^-\rangle|\Phi^-\rangle\langle\Phi^-|\langle\Phi^-|
\end{aligned}
$$

To see the effect of the CNOT gates, consider several examples. Consider the first term:

$$(CNOT_{A_1A_2} \otimes CNOT_{B_1B_2})|\Psi^-\rangle_{A_1B_1}|\Psi^-\rangle_{A_2B_2}$$

$$= (CNOT_{A_1A_2} \otimes CNOT_{B_1B_2})\frac{1}{2}(|0101\rangle_{A_1B_1A_2B_2} - |0110\rangle_{A_1B_1A_2B_2}$$

$$- |1001\rangle_{A_1B_1A_2B_2} + |1010\rangle_{A_1B_1A_2B_2})$$

$$= \frac{1}{2}(|0100\rangle_{A_1B_1A_2B_2} - |0111\rangle_{A_1B_1A_2B_2} - |1011\rangle_{A_1B_1A_2B_2} + |1000\rangle_{A_1B_1A_2B_2})$$

$$= |\Psi^+\rangle_{A_1B_1}|\Phi^-\rangle_{A_2B_2}$$

Consider the second term:

$$(CNOT_{A_1A_2} \otimes CNOT_{B_1B_2})|\Psi^-\rangle_{A_1B_1}|\Psi^+\rangle_{A_2B_2}$$

$$= (CNOT_{A_1A_2} \otimes CNOT_{B_1B_2})\frac{1}{2}(|0101\rangle_{A_1B_1A_2B_2} + |0110\rangle_{A_1B_1A_2B_2}$$

$$- |1001\rangle_{A_1B_1A_2B_2} - |1010\rangle_{A_1B_1A_2B_2})$$

$$= \frac{1}{2}(|0100\rangle_{A_1B_1A_2B_2} + |0111\rangle_{A_1B_1A_2B_2} - |1011\rangle_{A_1B_1A_2B_2} - |1000\rangle_{A_1B_1A_2B_2})$$

$$= |\Psi^-\rangle_{A_1B_1}|\Phi^+\rangle_{A_2B_2}$$

Consider the third term:

$$(CNOT_{A_1A_2} \otimes CNOT_{B_1B_2})|\Psi^-\rangle_{A_1B_1}|\Phi^+\rangle_{A_2B_2}$$

$$= (CNOT_{A_1A_2} \otimes CNOT_{B_1B_2})\frac{1}{2}(|0100\rangle_{A_1B_1A_2B_2} + |0111\rangle_{A_1B_1A_2B_2}$$

$$- |1000\rangle_{A_1B_1A_2B_2} - |1011\rangle_{A_1B_1A_2B_2})$$

$$= \frac{1}{2}(|0101\rangle_{A_1B_1A_2B_2} + |0110\rangle_{A_1B_1A_2B_2} - |1010\rangle_{A_1B_1A_2B_2} - |1001\rangle_{A_1B_1A_2B_2})$$

$$= |\Psi^-\rangle_{A_1B_1}|\Psi^+\rangle_{A_2B_2}$$

In other words, we can see that the effect of applying the CNOT gates is to map from a composition of two Bell states to another composition of two Bell states.

We only saw calculations for three examples, but we can consider all the possible terms and tabulate the probabilities after step 3, as in Bennett [1996], the first four columns with entries showing only the symbol for the Bell states, as shown in the table below. We then do a measurement in Pauli Z on the target qubits (A_2B_2) and keep only the source (control) qubits (A_1B_1) for when both measurement results of the target qubits are equal, i.e. only the cases where (A_2B_2) is either the $|\Phi^+\rangle$ state or the $|\Phi^-\rangle$ state; the probability of each such case is highlighted in the table (the probabilities of the other cases are "discarded").

Before CNOT gates		After CNOT gates		Probability	$\sigma''_{A_1B_1}$
A_1B_1	A_2B_2 (targ)	A_1B_1	A_2B_2 (targ)		
Φ^+	Φ^+	Φ^+	Φ^+	$c_{11} = F^2$	Keep
Φ^-	Φ^+	Φ^-	Φ^+	$c_{15} = \frac{(1-F)}{3} \cdot F$	Keep
Ψ^+	Φ^+	Ψ^+	Ψ^+	$c_7 = \frac{(1-F)}{3} \cdot F$	Discarded
Ψ^-	Φ^+	Ψ^-	Ψ^+	$c_3 = \frac{(1-F)}{3} \cdot F$	Discarded
Ψ^+	Ψ^+	Ψ^+	Φ^+	$c_6 = \frac{(1-F)}{3} \cdot \frac{(1-F)}{3}$	Keep
Ψ^-	Ψ^+	Ψ^-	Φ^+	$c_2 = \frac{(1-F)}{3} \cdot \frac{(1-F)}{3}$	Keep
Φ^+	Ψ^+	Φ^+	Ψ^+	$c_{10} = F \cdot \frac{(1-F)}{3}$	Discarded
Φ^-	Ψ^+	Φ^-	Ψ^+	$c_{14} = \frac{(1-F)}{3} \cdot \frac{(1-F)}{3}$	Discarded
Φ^+	Φ^-	Φ^-	Φ^-	$c_{12} = F \cdot \frac{(1-F)}{3}$	Keep
Φ^-	Φ^-	Φ^+	Φ^-	$c_{16} = \frac{(1-F)}{3} \cdot \frac{(1-F)}{3}$	Keep
Ψ^+	Φ^-	Ψ^-	Ψ^-	$c_8 = \frac{(1-F)}{3} \cdot \frac{(1-F)}{3}$	Discarded
Ψ^-	Φ^-	Ψ^+	Ψ^-	$c_4 = \frac{(1-F)}{3} \cdot \frac{(1-F)}{3}$	Discarded
Ψ^+	Ψ^-	Ψ^-	Φ^-	$c_5 = \frac{(1-F)}{3} \cdot \frac{(1-F)}{3}$	Keep
Ψ^-	Ψ^-	Ψ^+	Φ^-	$c_1 = \frac{(1-F)}{3} \cdot \frac{(1-F)}{3}$	Keep
Φ^+	Ψ^-	Φ^-	Ψ^-	$c_9 = F \cdot \frac{(1-F)}{3}$	Discarded
Φ^-	Ψ^-	Φ^+	Ψ^-	$c_{13} = \frac{(1-F)}{3} \cdot \frac{(1-F)}{3}$	Discarded

After the CNOT gates, the probability F' of the source qubits (A_1B_1) being in the $|\Phi^+\rangle$ state when (A_2B_2) is $|\Phi^+\rangle$ or $|\Phi^-\rangle$ (highlighted rows in the table) is then:

$$F'F' = \frac{c_{11} + c_{16}}{c_{11} + (c_{15} + c_{12}) + (c_6 + c_2 + c_{16} + c_5 + c_1)} = \frac{F^2 + \frac{1}{9}(1-F)^2}{F^2 + \frac{2}{3}F(1-F) + \frac{5}{9}(1-F)^2}$$

where the denominator is the sum of the probabilities in the table where we got a $|\Phi^+\rangle$ state or the $|\Phi^-\rangle$ state in the target qubits (A_2B_2) (i.e. where both measurement results of the target qubits are equal, labeled "keep" in the last column), and the numerator is the sum of the probabilities where we got the $|\Phi^+\rangle$ state in the source qubits. The probability of getting two qubits the same (i.e. of keeping the source) is at least F^2, i.e. greater than $\frac{1}{4}$, if we assume $F > \frac{1}{2}$.

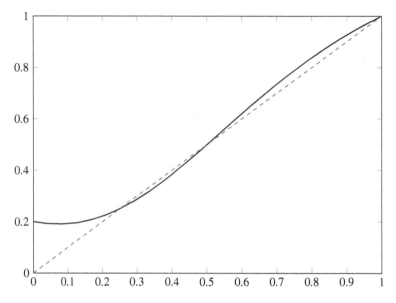

Figure 6.3 Graph of $F' = \dfrac{F^2 + \frac{1}{9}(1-F)^2}{F^2 + \frac{2}{3}F(1-F) + \frac{5}{9}(1-F)^2}$ in a plot of F' (vertical axis) against F (horizontal axis) showing that for $0.5 < F < 1$, $F' > F$. Dashed line is the linear graph of $F' = F$.

We note that if the two initial pairs $\rho_{A_1 B_1}$ and $\rho_{A_2 B_2}$ each have fidelity $F > \frac{1}{2}$, then the fidelity of the final pair $F' > F$; to see this, look at the graph in Figure 6.3. Note that if the CNOT gates were not performed, but the measurements and the filter done (to stop when $a \neq b$, i.e. continuing only when the results show the $|\Phi^+\rangle$ state or the $|\Phi^-\rangle$ state in the target qubits), we get the following (from rows 1–4, and 9–12):

$$F' = \frac{F^2 + \frac{1}{3}F(1 - F)}{F^2 + \frac{4}{3}F(1 - F) + \frac{3}{9}(1 - F)^2} = F$$

which does not help improve the fidelity. The CNOT operations help to change the states (or "move the probabilities") so that indeed $F' > F$, for $F' \geq 0.5$.

Finally, step 4 "moves the probabilities" back from the $|\Phi^+\rangle$ state to the singlet state (effectively reversing step 2) so that the probability of the quantum state being in the singlet state is F'. In summary, we have gone from (at least) two pairs of qubits, each prepared in a state of the form:

$$\sigma_{A_i B_i} = F |\Psi^-\rangle \langle \Psi^-| + \cdots$$

to a pair of qubits in a state of the form:

$$\sigma''_{A_i B_i} = F' \left| \Psi^- \right\rangle \left\langle \Psi^- \right| + \cdots$$

We can repeat this process – from at least one pair of states of fidelity $F > \frac{1}{2}$, go through the process to get a pair of fidelity $F' > F$ (we say "at least" since the process might fail to get a pair of higher fidelity, if the two measured qubits don't fulfill the condition, so that it might take more than one pair of qubits to get one higher fidelity pair), and then repeat this process to get a pair of fidelity $F'' > F'$, and so on, until F_{output} approaches 1.

With the above approach, we used two impure pairs of fidelity at least 0.5 to get one pair of higher fidelity. And we have to repeat this process many times for F_{output} to approach 1, throwing away at least half of the impure pairs (i.e. the targets). Hence, it is not a very efficient process. There are also concerns about what happens if two copies are not exact copies of each other and so on, which we will not go into here.

We have provided an outline of one method – there are further discussions on entanglement purification (or entanglement distillation) including other methods in Bennett et al. [1996], Deutsch et al. [1996], and Dür and Briegel [2007]. We discussed only bipartite Bell pairs above, but distillation protocols for multipartite *GHZ* states are described in de Bone et al. [2020].

6.3 Distributed Quantum Computation Over the Quantum Internet – Challenges

We can imagine an Internet of quantum repeaters which can be doing the tasks of entanglement swapping, or decoding and encoding qubits, e.g. in tree-cluster states, or other encodings. This means that suppose two nodes need to transmit states between each other or share some entangled EPR pair, then they can request for such entangled pairs from the quantum Internet. As a simple example, imagine that we have a three-qubit circuit as follows:

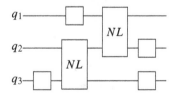

where qubit q_i is to execute on node QCi, respectively. We have two nonlocal gates, between QC1 and QC2 and between QC2 and QC3, marked *NL*. As we have seen

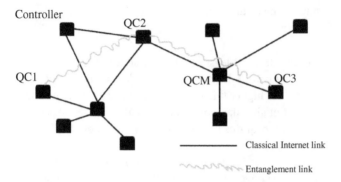

Figure 6.4 An illustration of three nodes QC1, QC2, and QC3, involved in a distributed quantum computation, with entanglement between qubits at QC1 and QC2 and between qubits at QC2 and QC3.

in Chapter 4, the nonlocal gates require entangled EPR pairs. A controller node takes the above circuit and assigns each node its corresponding computations, and coordinates their execution, collecting results thereafter. The controller node is connected to QC1, QC2, and QC3 via the classical Internet. Figure 6.4 illustrates this scenario.

One can see that there are timing constraints where nodes might need to wait for entanglement to be established, e.g. suppose between nodes which are far apart, or entanglement could be established *a priori*.[13] Also, one can imagine a much larger circuit where each node might be assigned multiple qubits, and a much more complex "Web" of entanglement between nodes.

Hence, acquiring entangled pairs between qubits on two nodes, for the purposes of nonlocal gates between the two nodes (as we have seen in Chapter 4), or for quantum teleportation, is a key challenge in distributed quantum computing over the quantum Internet. When two nodes do not yet share entangled pairs, entanglement swapping operations involving intermediate nodes (between the two nodes) might be needed before they can share entangled pairs, and so, perform distributed quantum computations. Hence, entangled pairs between nodes become an important resource to acquire.

We mentioned the idea of compilers for distributed quantum computing in Chapter 4. For a given quantum circuit to be executed over nodes connected over the classical and quantum Internet, there could be multiple ways each requiring different amount of resources, and so there is a need to consider different ways and choose the least resource-intensive alternative. For example, consider the

---

13 In principle, two nodes, each having a qubit of a shared entangled pair, could even move further apart (e.g. imagine QC1 and QC2 moving) and as long as they maintain those entangled qubits, they can compute what is required.

following quantum circuit C to be computed using nodes QC2 and QC3 connected as in Figure 6.4 via an intermediate node (call this QCM):

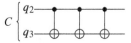

where q_i is located on QCi so that the three CNOT gates are actually distributed-CNOT operations. For each distributed-CNOT to be performed separately, an entangled pair involving (communication) qubits on the two nodes is needed as in Figure 4.2, here depicted as in Figure 6.5.

We can consider the resource requirements, e.g. in terms of the number of entangled pairs required, and time (i.e. the number of steps required to execute the circuit). One way to execute the quantum circuit is to perform each of the three distributed-CNOT gates separately. Now for each distributed-CNOT gate, an entangled pair is required – what resources are required for this? Assuming that an entangled pair is generated between QC2 and QCM and another entangled pair generated between QCM and QC3 (call these *neighbor entanglement links*). Then, an entanglement swapping operation is performed to get the entangled pair between QC2 and QC3, i.e. for each distributed-CNOT gate, 2 neighbor entanglement links are consumed, and suppose s steps of a quantum circuit are required for this, and then another d steps are required for the circuit in Figure 6.5, i.e. a total of $(s + d)$ steps are required for each distributed-CNOT gate between q_2 and q_3. For the quantum circuit C above with three distributed-CNOT operations, 6 neighbor entanglement links are consumed, and $3(s + d)$ steps are required. In general, suppose there are n distributed CNOTs to perform in a row instead of just 3, then $2n$ neighbor entanglement links are consumed and $n(s + d)$ steps are required.

Is there a way to reduce resources required? Indeed, there is, given that the control qubit is actually used three times without change in the circuit. One way

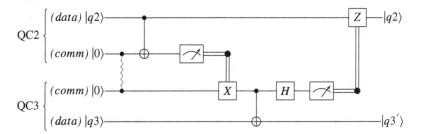

Figure 6.5 A circuit implementing a distributed-CNOT between qubits $q2$ and $q3$ across nodes QC2 and QC3. The communication qubits sharing an entangled pair are marked *comm*, and the two data qubits are marked *data*.

to reduce resources is to use a different approach, depicted in Figure 4.1, where a distributed CNOT is performed by teleporting the control qubit to the node with the target qubit, i.e. in this case, the control qubit q_2 can be teleported to QC3 and then used multiple times at QC3 (for the three CNOT operations) before being teleported back to QC2. If the number of steps required to create the entangled pair (via entanglement swapping) between QC2 and QC3 required for the teleportation is s steps (and 2 neighbor entanglement links are consumed), t steps are required to perform the teleportation of $|q_2\rangle$ from QC2 to QC3, three steps to do the three CNOT operations (locally) at QC3, and then another s steps to generate the entangled pair between QC2 and QC3 for the teleportation back (with another 2 neighbor entanglement links consumed for the entanglement swapping), and finally, t steps for the teleportation of $|q_2\rangle$ back from QC3 to QC2, then the number of steps required using this approach is $(s + t + 3 + s + t) = 2s + 2t + 3$, and $2 + 2 = 4$ neighbor entanglement links are consumed. In general, suppose there are n distributed CNOTs to perform instead of just 3, then $(s + t + n + s + t) = 2s + 2t + n$ steps are required, and still only 4 neighbor entanglement links are consumed. We note here that for large enough n (and $|t − d|$ a relatively small constant), $n(s + d) > 2s + 2t + n$, i.e. the second approach will be more efficient (using fewer number of steps) and requiring fewer neighbor entanglement links! Hence, a compiler should compile the (high-level) quantum circuit C into a (low-level) quantum circuit reflecting the second approach (when detecting a large enough n). Such compiler issues have been discussed more comprehensively and in detail in Ferrari et al. [2021] and Wu et al. [2022].

There are also issues of quantum routing where, for example, there could be multiple paths to form entanglement links between two nodes, with different possible performance characteristics. Allocation of resources in quantum networks to support distributed quantum computing applications is discussed in Cicconetti et al. [2022].

The quantum network and internetwork architecture are still being researched and developed. Corresponding to the current classical Internet architecture which uses layers of protocols, a similar layered protocol stack is being developed for the quantum Internet. A detailed discussion of current proposals for the quantum Internet protocol stack is given in Illiano et al. [2022], and there are standardization efforts (as mentioned in Chapter 1).[14] Hardware/device requirements for distributed quantum computing, including entanglement swapping, quantum memory, photon detectors, and so on, in the context of space quantum communications, have been identified.[15] An experimental demonstration of entanglement using a quantum network stack (including layers to generate

---

14 See https://irtf.org/qirg [last accessed: 16/10/2022].
15 See https://www.nasa.gov/sites/default/files/atoms/files/2020_quantum_workshop_report
.pdf [last accessed: 9/3/2023].

direct entanglement links (or neighbor entanglement links) and for end-to-end entanglement using entanglement swapping with the direct entanglement links) is described in Pompili et al. [2022].

A lot of the quantum protocols we have seen requires classical communications, e.g. quantum teleportation, but we have also seen superdense coding in Chapter 2, where we can (efficiently) transmit classical information using quantum information. The relationship between the quantum Internet and the classical Internet can be mutually beneficial; a review on how the quantum Internet can enhance classical Internet services is given by Cacciapuoti et al. [2022].

A topic which we did not discuss much about is distributed quantum sensing, which involves measurements on the quantum states resulting from applying unitary operations parameterized by the quantity being measured to an initial (reference) state – the reader is referred to Zhang and Zhuang [2021] for further reading. One can envision distributed quantum sensing integrated with distributed quantum computations.

There is an implementation of a number of distributed quantum computing algorithms using a simulation tool called QuNetSim [Diadamo et al., 2021],[16] and quantum networking protocols can be examined via the simulation tool called NetSquid [Coopmans et al., 2021].[17]

6.4 Summary

We have looked at how entanglement could be performed over longer distances and the notion of quantum repeaters to increase distances over which qubits might be teleported and over which qubits might be entangled. A network of such quantum repeaters can support nodes wanting to exchange quantum information or to perform nonlocal gates, even if they are far apart. We have also looked at one way to increase the fidelity of entangled pairs (though there are others), and we then briefly saw how distributed quantum computing happens on over the Internet, including challenges.

References

Charles H. Bennett, Gilles Brassard, Sandu Popescu, Benjamin Schumacher, John A. Smolin, and William K. Wootters. Purification of noisy entanglement and faithful teleportation via noisy channels. *Physical Review Letters*, 76:722–725, Jan 1996. doi: 10.1103/PhysRevLett.76.722.

---

16 https://tqsd.github.io/QuNetSim [last accessed: 22/3/2023].
17 https://netsquid.org [last accessed: 22/3/2023].

Sebastian de Bone, Runsheng Ouyang, Kenneth Goodenough, and David Elkouss. Protocols for creating and distilling multipartite GHZ states with bell pairs. *IEEE Transactions on Quantum Engineering*, 1:1–10, 2020. doi: 10.1109/TQE.2020 .3044179.

Johannes Borregaard, Hannes Pichler, Tim Schröder, Mikhail D. Lukin, Peter Lodahl, and Anders S. Sørensen. One-way quantum repeater based on near-deterministic photon-emitter interfaces. *Physical Review X*, 10:021071, Jun 2020. doi: 10.1103/PhysRevX.10.021071.

Angela Sara Cacciapuoti, Jessica Illiano, Seid Koudia, Kyrylo Simonov, and Marcello Caleffi. The quantum internet: enhancing classical internet services one qubit at a time. *IEEE Network*, 36(5):6–12, 2022. doi: 10.1109/MNET.001.2200162.

C. Cicconetti, M. Conti, and A. Passarella. Resource allocation in quantum networks for distributed quantum computing. In *2022 IEEE International Conference on Smart Computing (SMARTCOMP)*, pages 124–132, Los Alamitos, CA, USA, Jun 2022. IEEE Computer Society. doi: 10.1109/SMARTCOMP55677.2022.00032. URL https://doi.ieeecomputersociety.org/10.1109/SMARTCOMP55677.2022.00032.

Tim Coopmans, Robert Knegjens, Axel Dahlberg, David Maier, Loek Nijsten, Julio de Oliveira Filho, Martijn Papendrecht, Julian Rabbie, Filip Rozpedek, Matthew Skrzypczyk, Leon Wubben, Walter de Jong, Damian Podareanu, Ariana Torres-Knoop, David Elkouss, and Stephanie Wehner. NetSquid, a network simulator for quantum information using discrete events. *Communications Physics*, 4(1):164, 2021.

David Deutsch, Artur Ekert, Richard Jozsa, Chiara Macchiavello, Sandu Popescu, and Anna Sanpera. Quantum privacy amplification and the security of quantum cryptography over noisy channels. *Physical Review Letters*, 77:2818–2821, Sep 1996. doi: 10.1103/PhysRevLett.77.2818.

Stephen Diadamo, Janis Nötzel, Benjamin Zanger, and Mehmet Mert Beşe. QuNetSim: a software framework for quantum networks. *IEEE Transactions on Quantum Engineering*, 2:1–12, 2021. doi: 10.1109/TQE.2021.3092395.

W. Dür and H. J. Briegel. Entanglement purification and quantum error correction. *Reports on Progress in Physics*, 70(8):1381, Jul 2007. doi: 10.1088/0034-4885/70/8/R03.

Davide Ferrari, Angela Sara Cacciapuoti, Michele Amoretti, and Marcello Caleffi. Compiler design for distributed quantum computing. *IEEE Transactions on Quantum Engineering*, 2:1–20, 2021. doi: 10.1109/TQE.2021.3053921.

Jessica Illiano, Marcello Caleffi, Antonio Manzalini, and Angela Sara Cacciapuoti. Quantum internet protocol stack: a comprehensive survey. *Computer Networks*, 213:109092, 2022.

Sreraman Muralidharan, Linshu Li, Jungsang Kim, Norbert Lütkenhaus, Mikhail D. Lukin, and Liang Jiang. Optimal architectures for long distance quantum communication. *Scientific Reports*, 6(1):20463, 2016.

M. Pompili, C. Delle Donne, I. te Raa, B. van der Vecht, M. Skrzypczyk, G. Ferreira, L. de Kluijver, A. J. Stolk, S. L. N. Hermans, P. Pawełczak, W. Kozlowski, R. Hanson, and S. Wehner. Experimental demonstration of entanglement delivery using a quantum network stack. *npj Quantum Information*, 8(1):121, 2022.

Mathias Soeken, D. Michael Miller, and Rolf Drechsler. Quantum circuits employing roots of the pauli matrices. *Physical Review A*, 88:042322, Oct 2013. doi: 10.1103/PhysRevA.88.042322.

Anbang Wu, Hezi Zhang, Gushu Li, Alireza Shabani, Yuan Xie, and Yufei Ding. AutoComm: a framework for enabling efficient communication in distributed quantum programs. In *2022 55th IEEE/ACM International Symposium on Microarchitecture (MICRO)*, pages 1027–1041, 2022. doi: 10.1109/MICRO56248.2022.00074.

Pei-Shun Yan, Lan Zhou, Wei Zhong, and Yu-Bo Sheng. A survey on advances of quantum repeater. *Europhysics Letters*, 136(1):14001, Nov 2021.

Zheshen Zhang and Quntao Zhuang. Distributed quantum sensing. *Quantum Science and Technology*, 6(4):043001, Jul 2021.

7

Conclusion

We have seen that entanglement and teleportation are central concepts for distributed quantum computing. With "small" quantum networks (and there are already examples [Dowling, 2020; Fürnkranz, 2020]), there is potential for distributed quantum computing over such networks, at least initially. The quantum Internet is still nascent, but as it develops, it will allow entanglement and teleportation over longer distances in a reliable way, enabling nodes to exchange quantum information and work together quantumly across the Internet, beyond just working together classically today. Distributed quantum computing will likely happen over such small networks first and then scale up as networks get interconnected and the quantum Internet emerges.

Greater software abstraction and software support will be needed including tools to support distributed quantum computing and to work with possibly a wide range of quantum hardware. Quantum hardware itself is an area of active development, with different quantum hardware architectures in experimentation, not just for local computations, but for computations involving entanglement and teleportation. Also, relevant theory for quantum communications and the quantum Internet is progressing.[1] Hence, there are developments in theory, practice, applications, hardware, and software, as the field progresses.

This book is perhaps just the tip of the iceberg of a selection of topics, and there is much work still to be done, but it is hoped that it can provide newcomers a "first taste" of what we have called *Quantum Internet Computing*.

---

1 See the book on "Principles of Quantum Communication Theory: A Modern Approach" at https://arxiv.org/pdf/2011.04672.pdf [last accessed: 21/11/2022].

From Distributed Quantum Computing to Quantum Internet Computing: An Introduction, First Edition. Seng W. Loke.
© 2024 The Institute of Electrical and Electronics Engineers, Inc. Published 2024 by John Wiley & Sons, Inc.

References

Jonathan P. Dowling. *Schrödinger's Web: Race to Build the Quantum Internet.* CRC Press, 2020.

Gösta Fürnkranz. *The Quantum Internet: Ultrafast and Safe from Hackers.* Springer, 2020.

Index

From Distributed Quantum Computing to Quantum Internet Computing: An Introduction,
First Edition. Seng W. Loke.
© 2024 The Institute of Electrical and Electronics Engineers, Inc. Published 2024 by John Wiley & Sons, Inc.

Printed in the USA
CPSIA information can be obtained
at www.ICGtesting.com
CBHW070752290524
9015CB00051B/9

9 781394 185511